Selected Titles in This Series

716 **John H. Palmieri,** Stable homotopy over the Steenrod algebra, 2001

715 **W. N. Everitt and L. Markus,** Multi-interval linear ordinary boundary value problems and complex symplectic algebra, 2001

714 **Earl Berkson, Jean Bourgain, and Aleksander Pełczynski,** Canonical Sobolev projections of weak type $(1,1)$, 2001

713 **Dorina Mitrea, Marius Mitrea, and Michael Taylor,** Layer potentials, the Hodge Laplacian, and global boundary problems in nonsmooth Riemannian manifolds, 2001

712 **Raúl E. Curto and Woo Young Lee,** Joint hyponormality of Toeplitz pairs, 2001

711 **V. G. Kac, C. Martinez, and E. Zelmanov,** Graded simple Jordan superalgebras of growth one, 2001

710 **Brian Marcus and Selim Tuncel,** Resolving Markov chains onto Bernoulli shifts via positive polynomials, 2001

709 **B. V. Rajarama Bhat,** Cocylces of CCR flows, 2001

708 **William M. Kantor and Ákos Seress,** Black box classical groups, 2001

707 **Henning Krause,** The spectrum of a module category, 2001

706 **Jonathan Brundan, Richard Dipper, and Alexander Kleshchev,** Quantum Linear groups and representations of $GL_n(\mathbb{F}_q)$, 2001

705 **I. Moerdijk and J. J. C. Vermeulen,** Proper maps of toposes, 2000

704 **Jeff Hooper, Victor Snaith, and Min van Tran,** The second Chinburg conjecture for quaternion fields, 2000

703 **Erik Guentner, Nigel Higson, and Jody Trout,** Equivariant E-theory for C^*-algebras, 2000

702 **Ilijas Farah,** Analytic guotients: Theory of liftings for quotients over analytic ideals on the integers, 2000

701 **Paul Selick and Jie Wu,** On natural coalgebra decompositions of tensor algebras and loop suspensions, 2000

700 **Vicente Cortés,** A new construction of homogeneous quaternionic manifolds and related geometric structures, 2000

699 **Alexander Fel'shtyn,** Dynamical zeta functions, Nielsen theory and Reidemeister torsion, 2000

698 **Andrew R. Kustin,** Complexes associated to two vectors and a rectangular matrix, 2000

697 **Deguang Han and David R. Larson,** Frames, bases and group representations, 2000

696 **Donald J. Estep, Mats G. Larson, and Roy D. Williams,** Estimating the error of numerical solutions of systems of reaction-diffusion equations, 2000

695 **Vitaly Bergelson and Randall McCutcheon,** An ergodic IP polynomial Szemerédi theorem, 2000

694 **Alberto Bressan, Graziano Crasta, and Benedetto Piccoli,** Well-posedness of the Cauchy problem for $n \times n$ systems of conservation laws, 2000

693 **Doug Pickrell,** Invariant measures for unitary groups associated to Kac-Moody Lie algebras, 2000

692 **Mara D. Neusel,** Inverse invariant theory and Steenrod operations, 2000

691 **Bruce Hughes and Stratos Prassidis,** Control and relaxation over the circle, 2000

690 **Robert Rumely, Chi Fong Lau, and Robert Varley,** Existence of the sectional capacity, 2000

689 **M. A. Dickmann and F. Miraglia,** Special groups: Boolean-theoretic methods in the theory of quadratic forms, 2000

688 **Piotr Hajłasz and Pekka Koskela,** Sobolev met Poincaré, 2000

687 **Guy David and Stephen Semmes,** Uniform rectifiability and quasiminimizing sets of arbitrary codimension, 2000

(Continued in the back of this publication)

Stable Homotopy over the Steenrod Algebra

of the
American Mathematical Society

Number 716

Stable Homotopy over
the Steenrod Algebra

John H. Palmieri

May 2001 • Volume 151 • Number 716 (second of 5 numbers) • ISSN 0065-9266

American Mathematical Society
Providence, Rhode Island

2000 *Mathematics Subject Classification.*
Primary 55S10, 55U15, 18G35, 55U35, 55T15, 55P42, 55Q10, 55Q45, 18G15, 16W30, 18E30, 20J99.

Library of Congress Cataloging-in-Publication Data

Palmieri, John H. (John Harold), 1964–
  Stable homotopy over the Steenrod algebra / John H. Palmieri.
    p. cm. — (Memoirs of the American Mathematical Society, ISSN 0065-9266 ; no. 716)
  "May 2001, volume 151, number 716 (second of 5 numbers)."
  Includes bibliographical references and index.
  ISBN 0-8218-2668-9 (alk. paper)
  1. Homotopy theory. 2. Steenrod algebra. I. Title. II. Series.
QA3.A57 no. 716
[QA612.7]
510 s—dc21
[514′.24]   2001018228

Memoirs of the American Mathematical Society

This journal is devoted entirely to research in pure and applied mathematics.

Subscription information. The 2001 subscription begins with volume 149 and consists of six mailings, each containing one or more numbers. Subscription prices for 2001 are $494 list, $395 institutional member. A late charge of 10% of the subscription price will be imposed on orders received from nonmembers after January 1 of the subscription year. Subscribers outside the United States and India must pay a postage surcharge of $31; subscribers in India must pay a postage surcharge of $43. Expedited delivery to destinations in North America $35; elsewhere $130. Each number may be ordered separately; *please specify number* when ordering an individual number. For prices and titles of recently released numbers, see the New Publications sections of the *Notices of the American Mathematical Society*.

Back number information. For back issues see the *AMS Catalog of Publications*.

Subscriptions and orders should be addressed to the American Mathematical Society, P. O. Box 845904, Boston, MA 02284-5904. *All orders must be accompanied by payment*. Other correspondence should be addressed to Box 6248, Providence, RI 02940-6248.

Copying and reprinting. Individual readers of this publication, and nonprofit libraries acting for them, are permitted to make fair use of the material, such as to copy a chapter for use in teaching or research. Permission is granted to quote brief passages from this publication in reviews, provided the customary acknowledgment of the source is given.

Republication, systematic copying, or multiple reproduction of any material in this publication is permitted only under license from the American Mathematical Society. Requests for such permission should be addressed to the Assistant to the Publisher, American Mathematical Society, P. O. Box 6248, Providence, Rhode Island 02940-6248. Requests can also be made by e-mail to reprint-permission@ams.org.

Memoirs of the American Mathematical Society is published bimonthly (each volume consisting usually of more than one number) by the American Mathematical Society at 201 Charles Street, Providence, RI 02904-2294. Periodicals postage paid at Providence, RI. Postmaster: Send address changes to Memoirs, American Mathematical Society, P. O. Box 6248, Providence, RI 02940-6248.

© 2001 by the American Mathematical Society. All rights reserved.
This publication is indexed in *Science Citation Index*®, *SciSearch*®, *Research Alert*®, *CompuMath Citation Index*®, *Current Contents*®/*Physical, Chemical & Earth Sciences*.
Printed in the United States of America.

∞ The paper used in this book is acid-free and falls within the guidelines established to ensure permanence and durability.
Visit the AMS home page at URL: http://www.ams.org/

10 9 8 7 6 5 4 3 2 1   06 05 04 03 02 01

Contents

List of Figures	x
Preface	xi
Chapter 0. Preliminaries	1
0.1. Grading and other conventions	1
0.2. Hopf algebras	2
0.3. Modules and comodules	4
0.4. Homological algebra	6
0.5. Two small examples	9
Chapter 1. Stable homotopy over a Hopf algebra	13
1.1. The category Stable(Γ)	14
1.2. The functor H	16
1.2.1. Remarks on Hopf algebra extensions	20
1.3. Some classical homotopy theory	22
1.4. The Adams spectral sequence	25
1.5. Bousfield classes and Brown-Comenetz duality	29
1.6. Further discussion	31
Chapter 2. Basic properties of the Steenrod algebra	35
2.1. Quotient Hopf algebras of A	35
2.1.1. Quasi-elementary quotients of A	41
2.2. P_t^s-homology	42
2.2.1. Miscellaneous results about P_t^s-homology	46
2.3. Vanishing lines for homotopy groups	48
2.3.1. Proof of Theorems 2.3.1 and 2.3.2 when $p = 2$	50
2.3.2. Changes necessary when p is odd	55
2.4. Self-maps via vanishing lines	57
2.5. Construction of spectra of specified type	60
2.6. Further discussion	63
Chapter 3. Chromatic structure	65
3.1. Margolis' killing construction	65
3.2. A Tate version of the functor H	71
3.3. Chromatic convergence	75
3.4. Further discussion: work of Mahowald and Shick	76
3.5. Further discussion	77
Chapter 4. Computing Ext with elements inverted	79
4.1. The q_n-based Adams spectral sequence	80

4.2.	The Q_n-based Adams spectral sequence	81
4.3.	$A(n)$ as an A-comodule	85
4.4.	$\frac{1}{2}A(n)$ satisfies the vanishing plane condition	88
4.5.	$\frac{1}{2}A(n)$ generates the expected thick subcategory	90
4.5.1.	The proof of Proposition 4.5.7	95
4.6.	Some computations and applications	99
4.6.1.	Computation of $(Q_n)_{**}(Q_n)$	99
4.6.2.	Eisen's calculation	102
4.6.3.	The v_1-inverted Ext of the mod 2 Moore spectrum	104

Chapter 5. Quillen stratification and nilpotence — 107
  5.1. Statements of theorems — 108
  5.1.1. Quillen stratification — 108
  5.1.2. Nilpotence — 110
  5.2. Nilpotence and F-isomorphism via the Hopf algebra D — 112
  5.2.1. Nilpotence: Proof of Theorem 5.1.5 — 116
  5.2.2. F-isomorphism: Proof of Theorem 5.1.2 — 117
  5.3. Nilpotence and F-isomorphism via quasi-elementary quotients — 119
  5.3.1. Nilpotence: Proof of Theorem 5.1.6 — 119
  5.3.2. F-isomorphism: Proof of Theorem 5.1.3 — 121
  5.4. Further discussion: nilpotence at odd primes — 123
  5.5. Further discussion: miscellany — 124

Chapter 6. Periodicity and other applications of the nilpotence theorems — 127
  6.1. The periodicity theorem — 127
  6.2. y-maps and their properties — 128
  6.3. Properties of ideals — 131
  6.4. The proof of the periodicity theorem — 132
  6.5. Computation of some invariants in HD_{**} — 134
  6.6. Computation of a few Bousfield classes — 139
  6.7. Ideals and thick subcategories — 143
  6.7.1. The thick subcategory conjecture — 143
  6.7.2. Rank varieties — 146
  6.8. Further discussion: slope supports — 147
  6.9. Further discussion: miscellany — 150

Appendix A. An underlying model category — 153

Appendix B. Steenrod operations and nilpotence in $\operatorname{Ext}_\Gamma^{**}(k,k)$ — 155
  B.1. Steenrod operations in Hopf algebra cohomology — 155
  B.2. Nilpotence in Ext over quotients of A: $p=2$ — 156
  B.3. Nilpotence in Ext over quotients of A: p odd — 157
  B.3.1. Sketch of proof of Conjecture B.3.4, and other results — 159

Appendix. Bibliography — 165

Appendix. Index — 169

ABSTRACT. We apply the tools of stable homotopy theory to the study of modules over the mod p Steenrod algebra A^*. More precisely, let A be the dual of A^*; then we study the category $\mathsf{Stable}(A)$ of unbounded cochain complexes of injective comodules over A, in which the morphisms are cochain homotopy classes of maps. This category is triangulated. Indeed, it is a stable homotopy category, so we can use Brown representability, Bousfield localization, Brown-Comenetz duality, and other homotopy-theoretic tools to study it. One focus of attention is the analogue of the stable homotopy groups of spheres, which in this setting is the cohomology of A, $\operatorname{Ext}_A^{*,*}(\mathbf{F}_p, \mathbf{F}_p)$. We also have nilpotence theorems, periodicity theorems, a convergent chromatic tower, and a number of other results.

Research partially supported by National Science Foundation grant DMS-9407459.

Key words and phrases. Steenrod algebra, Ext, stable homotopy, spectra, nilpotence theorem, Quillen stratification, chromatic convergence

List of Figures

2.1.A	Graphical representation of a quotient Hopf algebra of A.	37
2.1.B	Profile functions for $A(n)$.	38
2.1.C	Profile functions for maximal elementary quotients of A, $p=2$.	40
2.3.A	Vanishing line at the prime 2.	49
3.1.A	Vanishing curve for $\pi_{ij}(C_n^f S^0)$.	70
3.2.A	The coefficients of $HA(1)$ and $\widehat{H}A(1)$.	74
4.2.A	Vanishing plane in Theorem 4.2.6.	83
4.6.A	"Profile function" for $H(A \square_E \mathbf{F}_p, Q_n)$.	100
4.6.B	Lightning flashes and v_1-towers.	105
5.1.A	Profile functions for D.	108
5.2.A	Profile function for $D(n)$.	113
5.3.A	Profile functions for D_r and $D_{r,q}$.	120
6.5.A	Graphical depiction of coaction of A on $\lim HE_{**}$.	138
6.8.A	$T(t,s)$ and $T(m)$ as subsets of Slopes$'$.	149

Preface

The object of study for this book is the mod p Steenrod algebra A and its cohomology $\mathrm{Ext}_A^{**}(\mathbf{F}_p, \mathbf{F}_p)$. Various people, including the author, have approached this subject by taking results in stable homotopy theory and then trying to prove analogous results for A-modules. This has proven to be successful, but the analogies were just that—there was no formal setting in which to do anything more precise.

In [**HPS97**], Hovey, Strickland, and the author developed "axiomatic stable homotopy theory." In particular, we gave axioms for a *stable homotopy category*; in any such category, one has available many of the tools of classical and modern stable homotopy theory—tools like Brown representability and Bousfield localization. It turns out that a category Stable(A) (defined in the next paragraph) of modules over the Steenrod algebra is such a category; as one might expect, the trivial module $\mathbf{F}_p$ plays the role of the sphere spectrum S^0, and $\mathrm{Ext}_A^{**}(-,-)$ plays the role of homotopy classes of maps. Since many of the tools of stable homotopy theory are focused on the study of the homotopy groups of S^0 (and of other spectra), one should expect the corresponding tools in Stable(A) to help in the study of $\mathrm{Ext}_A^{**}(\mathbf{F}_p, \mathbf{F}_p)$ (and related groups). In this book we apply some of these tools—Adams spectral sequences, nilpotence theorems, periodicity theorems, chromatic towers, etc.—to the study of Ext over the Steenrod algebra. It is our hope that this book will serve two purposes: first, to provide a reference source for a number of results about the Steenrod algebra and its cohomology, and second, to provide an example of an in-depth use of the language and tools of axiomatic stable homotopy theory in an algebraic setting.

Now we describe the category in which we work. We fix a prime p, let A^* be the mod p Steenrod algebra, and let $A = \mathrm{Hom}_{\mathbf{F}_p}(A^*, \mathbf{F}_p)$ be the (graded) dual of the Steenrod algebra. We let Stable(A) be the category whose objects are unbounded cochain complexes of injective left A-comodules, and whose morphisms are cochain homotopy classes of maps. This is a stable homotopy category. We prove a number of results about Stable(A); some of these are analogues of results in the ordinary stable homotopy category, and some are not. Some of these are new, and some already known, at least in the setting of A^*-modules; the old results often need new proofs to apply in the more general setting we discuss here.

NOTE. This work arose from the study of the abelian category of left A^*-modules; to apply stable homotopy theoretic techniques, though, it is most convenient to work in a triangulated category. One's first guess for an appropriate category might be one with objects the chain complexes of projective A^*-modules; it turns out that this category has some technical difficulties—see Remark 1.1.1(d). It is much more convenient to work with A-comodules instead of A^*-modules, and fortunately, one does not lose much by doing this. Many A^*-modules of interest can be viewed as A-comodules; the main effects of using comodules are things of the following sort: various arrows go the "wrong" way, $\operatorname{Ext}_A^{**}(\mathbf{F}_p, \mathbf{F}_p)$ is covariant in A, and one studies A via its quotient Hopf algebras (because those are dual to the sub-Hopf algebras of A^*).

Each chapter is divided into a number of sections; at the beginning of each chapter, we give a brief description of its contents, section by section. In this introduction, we give a brief overview of each chapter. We note that most chapters have at least one "Further discussion" section, in which we discuss issues auxiliary to the general discussion.

In Chapter 0, we provide some background material. We introduce some conventions related to grading; then we discuss Hopf algebras, modules, and comodules. We define what it means for a comodule to be injective, and we use this to define Ext in the category of comodules. We end the chapter by looking at some examples of small Hopf algebras; we classify the indecomposable comodules over these, and we compute the Ext groups for these indecomposables.

In Chapter 1, given a graded commutative Hopf algebra Γ over a field k, we define and examine the category $\mathsf{Stable}(\Gamma)$. The main examples we are concerned with are when Γ is the dual of a group algebra, the dual of an enveloping algebra, or the dual of the Steenrod algebra. Aside from setting up notation for use throughout this book, the main topics of this chapter include construction of cellular and Postnikov towers, an examination of generalized Adams spectral sequences in $\mathsf{Stable}(\Gamma)$, and some remarks on Bousfield classes and Brown-Comenetz duality.

In Chapter 2 we specialize to the case in which p is a prime, $k = \mathbf{F}_p$ is the field with p elements, and A is the dual of the mod p Steenrod algebra. Recall from [**Mil58**] that as algebras, we have

$$A \cong \begin{cases} \mathbf{F}_2[\xi_1, \xi_2, \xi_3, \ldots], & \text{if } p = 2, \\ \mathbf{F}_p[\xi_1, \xi_2, \xi_3, \ldots] \otimes \Lambda[\tau_0, \tau_1, \tau_2, \ldots], & \text{if } p \text{ is odd.} \end{cases}$$

The coproduct Δ on A is determined by

$$\Delta(\xi_n) = \sum_{i=0}^n \xi_{n-i}^{p^i} \otimes \xi_i,$$

$$\Delta(\tau_n) = \sum_{i=0}^n \xi_{n-i}^{p^i} \otimes \tau_i + \tau_n \otimes 1,$$

where $\xi_0 = 1$. In this chapter, we discuss two tools with which to study $\mathsf{Stable}(A)$: quotient Hopf algebras of A and P_t^s-homology. We use these tools to prove several theorems. The first is a vanishing line theorem, which generalizes results of Anderson-Davis [**AD73**] and Miller-Wilkerson [**MW81**]: given conditions on the P_t^s-homology groups of X, then $\operatorname{Ext}_A^{i,j}(\mathbf{F}_p, X) = 0$ when $i > mj - c$, for some numbers m and c. The second is a "self-map" theorem, which first appeared for modules in [**Pal92**]: given a finite A-comodule M, we construct a non-nilpotent element of $\operatorname{Ext}_A^{**}(M, M)$ satisfying certain properties. We use this second theorem to construct well-behaved objects with vanishing lines of specified slopes; these are analogues in the category $\mathsf{Stable}(A)$ of generalized Toda $V(n)$'s. We use these objects in Chapter 3.

In Chapter 3 we consider Steenrod algebra analogues of chromatic theory and the functors L_n and L_n^f. The latter functor L_n^f turns out be more tractable; in fact, it is a generalization (from the module setting) of Margolis' killing construction

[**Mar83**, Chapter 21]. We show that $L_n \neq L_n^f$ if $n > 1$, at least at the prime 2. We compute L_n^f on some particular ring spectra, and show that, at least for these rings, it turns "group cohomology" into "Tate cohomology." We use this result to show that the chromatic tower constructed using the functors L_n^f converges for any finite object. (This is a variant of a theorem of Margolis [**Mar83**, Theorem 22.1].)

In Chapter 4, we discuss a spectral sequence for computing $v_n^{-1}\operatorname{Ext}_A^{**}(\mathbf{F}_p, M)$ for certain comodules M (those for which some power of v_n acts on $\operatorname{Ext}_A^{**}(\mathbf{F}_p, M)$). As applications, we describe $v_0^{-1}\operatorname{Ext}_A^{**}(\mathbf{F}_p, M)$ in terms of the Q_0-homology of M, we reproduce Eisen's calculation [**Eis87**], and we discuss Mahowald's conjectured calculation in [**Mah70**] of the v_1-inverted Ext of the mod 2 Moore spectrum. The spectral sequence is constructed as a localization of an Adams spectral sequence, so convergence is an issue, but one which can be dealt with. As part of the discussion of convergence, we analyze the comodule structure on the Hopf algebras $A(n)$, originally due to Mitchell [**Mit85**] and Smith [**Smi**]. Of particular interest is the subcomodule $\frac{1}{2}A(n)$ of $A(n)$.

NOTE. Although Chapters 3 and 4 can be read independently, they are philosophically linked, in that they both involve using nice objects which have vanishing P_t^s-homology groups for specified P_t^s's. In Chapter 3, we use the analogues of generalized Toda $V(n)$'s to give a construction of the functor L_n^f. In Chapter 4, we study the comodules $\frac{1}{2}A(n)$ in order to prove convergence of the localized Adams spectral sequence.

Let $p = 2$. In Chapter 5 we develop analogues in the category $\mathsf{Stable}(A)$ of the nilpotence theorem of Devinatz, Hopkins, and Smith [**DHS88**], as well as the stratification theorem of Quillen [**Qui71**]. In fact we give two nilpotence theorems: in one we describe a single ring object (like BP) that detects nilpotence; more precisely, there is a quotient Hopf algebra D of A so that, if M is a finite-dimensional A-comodule, an element $z \in \operatorname{Ext}_A^{**}(M, M)$ is nilpotent under Yoneda composition if and only if its restriction to $\operatorname{Ext}_D^{**}(M, M)$ is nilpotent. The second nilpotence theorem is similar, but uses a family of ring objects (somewhat like the Morava K-theories) to detect nilpotence. These are versions in $\mathsf{Stable}(A)$ of the nilpotence theorems of [**DHS88**] and [**HS98**]. We strengthen these results when studying $\operatorname{Ext}_A^{**}(\mathbf{F}_2, \mathbf{F}_2)$, by "identifying" the image of $\operatorname{Ext}_A^{**}(\mathbf{F}_2, \mathbf{F}_2) \to \operatorname{Ext}_D^{**}(\mathbf{F}_2, \mathbf{F}_2)$ (and similarly for the other nilpotence theorem). One can view this as an analogue of Quillen's theorem [**Qui71**, Theorem 6.2], which identifies the cohomology of a compact Lie group up to F-isomorphism.

Again, let $p = 2$. In Chapter 6, we discuss applications and illustrations of the theorems from the previous chapter. In ordinary stable homotopy theory, the nilpotence theorems lead to the periodicity theorem and the thick subcategory theorem (see [**Hop87**] and [**HS98**]); in our setting, things are a bit harder, so we get a weak version of a periodicity theorem, and only a conjecture as to a classification of the thick subcategories of finite objects in $\mathsf{Stable}(A)$. More precisely, if M is a finite-dimensional A-comodule, then we produce a number of central non-nilpotent elements in $\operatorname{Ext}_A^{**}(M, M)$ by using the "ideal of M": the kernel of $\operatorname{Ext}_D^{**}(\mathbf{F}_2, \mathbf{F}_2) \to \operatorname{Ext}_D^{**}(M, M)$. We study some properties of these ideals, and we conjecture that the thick subcategories of finite-dimensional comodules are determined by them. One of our analogues of Quillen's theorem says that the elements of the kernel of $\operatorname{Ext}_A^{**}(\mathbf{F}_2, \mathbf{F}_2) \to \operatorname{Ext}_D^{**}(\mathbf{F}_2, \mathbf{F}_2)$ are nilpotent, and it identifies the image. This identification is not explicit, so we discuss some examples of elements known to be in the image. We also imitate [**Rav84**] to show that the objects that detect nilpotence have strictly smaller Bousfield classes than the sphere.

We also have two appendices: In Appendix A, we describe a model category whose associated homotopy category is Stable(Γ), for any commutative Hopf algebra Γ; the results here are due to Hovey [**Hov99**]. In Appendix B, we discuss the nilpotence of certain classes in $\text{Ext}_A^{**}(\mathbf{F}_p, \mathbf{F}_p)$ when A is the dual of the Steenrod algebra. The results depend on the prime; when $p = 2$, we get results that we use in Chapter 5 to prove our nilpotence theorems. When p is odd, we have more conjectures than results, and understanding those conjectures would be an important step in proving the nilpotence theorems at odd primes.

In this book we have a mix of results: some are extensions of older results to the cochain complex setting, and some are new. For each older result, if the proof in the literature extends easily to our setting, then we do not include a proof; otherwise, we at least give a sketch. It appears that when one uses the language of stable homotopy theory, one tends to change arguments with spectral sequences into simpler arguments with cofibration sequences (see Lemma 1.2.15, for example), so even though the setting is potentially more complicated, or at least less familiar, some of the proofs simplify. In such cases, we often give in to temptation and include the new proof in its entirety (as, for example, with the vanishing line theorem 2.3.1). Obviously, we include full proofs of all of the new results, and we give references for all of the old results.

ACKNOWLEDGMENTS: I have had a number of entertaining and illuminating discussions with a number of people on this material, Mark Hovey perhaps more than most. Others include Haynes Miller, Mike Hopkins, Dan Christensen, Hal Sadofsky, Bill Dwyer, and Paul Shick.

The following chart indicates the interdependencies of the chapters. (This chart includes the primary connections but omits a few small ones; for instance, the proof of Theorem 6.8.3 uses a result from Chapter 4.)

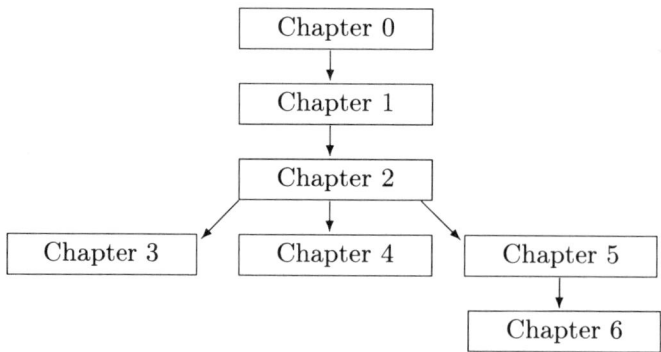

CHAPTER 0

Preliminaries

In this chapter, we give preliminaries and background. Almost all of the material here is standard, but it may be helpful to readers who are more familiar with modules than comodules.

We start in Section 0.1 by giving our grading conventions. In Section 0.2 we define Hopf algebras and mention a few examples. We define modules and comodules in the next section; in this book, we focus primarily on comodules, so we also discuss the relationship between comodules over a coalgebra and modules over the dual algebra. In Section 0.4 we discuss homological algebra in the category of comodules, recalling the definitions of injective objects and of Ext. In the last section of the chapter, we discuss in some detail two examples of small Hopf algebras, comodules over them, and the computation of their Ext groups.

0.1. Grading and other conventions

Fix a field k. We work throughout with **Z**-graded vector spaces over k; every element from such a vector space is assumed to be homogeneous, unless otherwise indicated. If V is a graded vector space, we write the ith graded piece as V_i, so that $V = \bigoplus_{i \in \mathbf{Z}} V_i$. V is of *finite type* if each V_i is finite-dimensional. V is *bounded below* if $V_i = 0$ for all sufficiently small i. Given a homogeneous element v of a graded vector space, we write $|v|$ for the *degree* of v. We write s for the "shift" or "suspension" functor on graded vector spaces: for a graded vector space $V = \bigoplus_{i \in \mathbf{Z}} V_i$, the ith graded piece of sV is $(sV)_i = V_{i-1}$. Of course, we may iterate s any number of times to define s^n for any $n \geq 0$; s is also invertible, so s^n makes sense for any integer n. We also use s for the induced suspension functor on the categories of graded modules and graded comodules. (Note that in Section 1.1 we introduce another, doubly-indexed, suspension functor $\Sigma^{i,j}$.)

A *degree zero* map between graded vector spaces sends homogeneous elements to homogeneous elements of the same degree; we write $\mathrm{Hom}^0(V, W)$ for the collection of such maps. For any integer j, the set of *degree j* maps is the set

$$\mathrm{Hom}^j(V, W) = \mathrm{Hom}^0(s^j V, W).$$

$\mathrm{Hom}^*(V, W)$ denotes the set of all graded maps from V to W, of all degrees.

Throughout, every map between graded vector spaces is assumed to be a graded map of possibly nonzero degree, and we will often omit the degree. In other words, if V and W are graded vector spaces, the notation $f \colon V \to W$ indicates a graded map, not necessarily of degree zero. If there is an explicit suspension, though, then the map is understood to have the specified degree: $f \colon s^n V \to W$ indicates a map of degree zero from $s^n V$ to W, which is the same as a degree n map from V to W.

We will omit the word "graded" from this point on; all vector spaces and maps are understood to be graded.

There are several other important conventions: first, all unadorned tensor products are over the ground field k. Second, unless otherwise specified, "module" means "left module," and similarly for "comodule." Third, all algebras are assumed to be associative and unital; similarly, all coalgebras are coassociative and counital.

0.2. Hopf algebras

We start this section with the definition of a Hopf algebra. This is standard; two references are [**MM65**] and [**Swe69**]. Note that we insist that our Hopf algebras have antipodes.

DEFINITION 0.2.1. Fix a field k. A *Hopf algebra* over k is a graded k-vector space $\Gamma = \bigoplus_{i \in \mathbf{Z}} \Gamma_i$ together with the following structure maps (all of which are graded maps of degree zero):

- a *unit* or *coaugmentation* map $\eta \colon k \to \Gamma$,
- a *multiplication* map $\mu \colon \Gamma \otimes_k \Gamma \to \Gamma$,
- a *counit* or *augmentation* map $\varepsilon \colon \Gamma \to k$,
- a *comultiplication* or *diagonal* map $\Delta \colon \Gamma \to \Gamma \otimes_k \Gamma$,
- and a *conjugation* or *antipode* map $\chi \colon \Gamma \to \Gamma$.

The maps η and μ give Γ the structure of an associative unital k-algebra—i.e., the following two diagrams commute):

$$\begin{array}{ccc} \Gamma \otimes \Gamma \otimes \Gamma & \xrightarrow{1 \otimes \mu} & \Gamma \otimes \Gamma \\ \mu \otimes 1 \downarrow & & \downarrow \mu \\ \Gamma \otimes \Gamma & \xrightarrow{\mu} & \Gamma, \end{array}$$

and

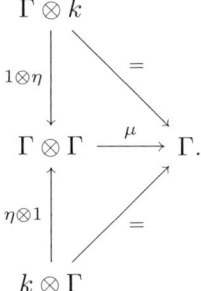

Dually, the maps ε and Δ make Γ into a coassociative counital k-coalgebra—the dual diagrams to those above (i.e., diagrams as above, but with all the arrows reversed) commute. We give $\Gamma \otimes \Gamma$ an algebra structure via the composite

$$\Gamma \otimes \Gamma \otimes \Gamma \otimes \Gamma \xrightarrow{1 \otimes T \otimes 1} \Gamma \otimes \Gamma \otimes \Gamma \otimes \Gamma \xrightarrow{\mu \otimes \mu} \Gamma \otimes \Gamma,$$

where $T \colon \Gamma \otimes \Gamma \to \Gamma \otimes \Gamma$ is the *twist map*: $T(a \otimes b) = (-1)^{|a||b|} b \otimes a$. (Dually, we can give $\Gamma \otimes \Gamma$ the structure of a coalgebra.) We insist that the maps Δ and ε be algebra maps; equivalently, we insist that μ and η be coalgebra maps. The conjugation map χ makes the following diagram commute:

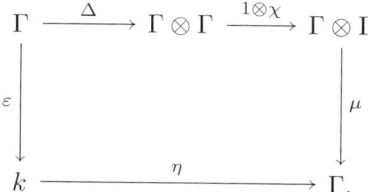

Lastly, the same diagram, except with $\chi \otimes 1$ replacing $1 \otimes \chi$, also commutes.

Since the structure maps in this definition are of degree zero, the image of the unit η lies in Γ_0, and the kernel of the counit ε contains $\bigoplus_{i \neq 0} \Gamma_i$. We also note that it is well-known—see [**Swe69**, Proposition 4.0.1], for example—that χ is an "anti-homomorphism" with respect to both the algebra and the coalgebra structures: $\chi \circ \mu = \mu \circ (\chi \otimes \chi) \circ T$, and $\Delta \circ \chi = T \circ (\chi \otimes \chi) \circ \Delta$.

DEFINITION 0.2.2. Let Γ be a Hopf algebra over a field k. We say that Γ is *commutative* if it is commutative as a graded algebra—i.e., the following diagram commutes:

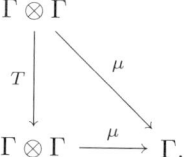

Γ is *cocommutative* if the dual diagram commutes; Γ is *bicommutative* if it is both commutative and cocommutative. We say that $\Gamma = \bigoplus_i \Gamma_i$ is *connected* if $\Gamma_i = 0$ when $i < 0$ and $\eta \colon k \to \Gamma_0$ is an isomorphism.

We work entirely with Hopf algebras that are either commutative (the usual situation) or cocommutative, or both. For such Hopf algebras, the antipode χ is an anti-automorphism of order 2: $\chi^2 = 1$. See [**Swe69**, Proposition 4.0.1].

Note that if Γ is a Hopf algebra for which each homogeneous piece Γ_i is finite-dimensional (i.e., Γ is of finite type), then the graded dual Γ^* of Γ has the structure of a Hopf algebra. (By "graded dual", we mean that $(\Gamma^*)_i = \mathrm{Hom}_k^i(\Gamma, k) = \mathrm{Hom}_k(\Gamma_{-i}, k)$.) In this case, quotient Hopf algebras of Γ correspond to sub-Hopf algebras of Γ^*, Γ is commutative if and only if Γ^* is cocommutative, etc. If some Γ_i is not finite-dimensional, then Γ^* will have the structure of an augmented algebra, and the dual of the product on Γ will make Γ^* a "completed" coalgebra rather than an honest coalgebra.

EXAMPLE 0.2.3. Let k be a field.
  (a) The homology of a topological group G with coefficients in k is a cocommutative Hopf algebra; it is connected if and only if G is connected.
  (b) For any group G, the group algebra kG is a cocommutative Hopf algebra. It is commutative if and only if G is abelian. It is graded trivially: every element is homogeneous of degree zero; hence it is connected if and only if G is the trivial group. If G is finite, then the vector space dual of kG is a commutative Hopf algebra.
  (c) For any Lie algebra L, its universal enveloping algebra $U(L)$ is a cocommutative Hopf algebra. It is commutative if and only if L is abelian. As with kG, it is graded trivially (unless L is graded itself).

(d) Similarly, if k has characteristic p, then for any restricted Lie algebra L defined over k, its restricted universal enveloping algebra $V(L)$ is a cocommutative Hopf algebra.

(e) Fix a prime p. The mod p Steenrod algebra A^* is a cocommutative Hopf algebra over the field $\mathbf{F}_p$; its dual A is a commutative Hopf algebra. It is conventional to grade each of these non-negatively, so that they are both connected; however, it is better to grade one positively and one negatively, so that they are graded duals of each other. We will grade A positively and use the induced negative grading on A^*. Starting in Chapter 2, we will focus almost exclusively on this example.

DEFINITION 0.2.4. An element γ of a Hopf algebra Γ is *primitive* if $\Delta(\gamma) = \gamma \otimes 1 + 1 \otimes \gamma$. We write $P\Gamma$ for the vector space of all primitives of Γ. An element γ is *grouplike* if $\Delta(\gamma) = \gamma \otimes \gamma$.

Note that a grouplike element must have degree zero, because the diagonal map Δ is a degree zero map.

0.3. Modules and comodules

We move on to a brief discussion of modules and comodules. Again, [**MM65**] is one of the standard references; [**Boa**] is also quite useful.

DEFINITION 0.3.1. Let Γ be a coalgebra over a field k. A k-vector space M is a (left) Γ-*comodule* if there is a degree zero structure map $\psi \colon M \to \Gamma \otimes M$, called the *coaction* map, making the following diagrams commute:

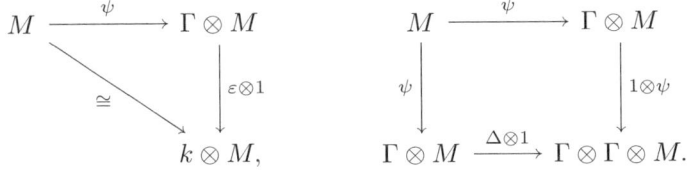

In other words, ψ defines a "coassociative" coaction. A (left) *module* over a k-algebra is defined dually, of course.

As mentioned in Section 0.1, we use left comodules and left modules throughout; from here on, we will omit the word "left."

Given a coalgebra map $\phi \colon \Gamma \to \Lambda$, any Γ-comodule M inherits a Λ-comodule structure, via the coaction map

$$M \to \Gamma \otimes M \xrightarrow{\phi \otimes 1} \Lambda \otimes M.$$

We call the Λ-comodule structure on M the "corestriction," the "push-forward," or (abusing the language a bit) the "restriction" of the Γ-comodule structure.

Given two Γ-comodules M and N over a Hopf algebra Γ, then $M \otimes N$ is naturally a Γ-comodule, via the structure map

$$M \otimes N \xrightarrow{\psi_M \otimes \psi_N} \Gamma \otimes M \otimes \Gamma \otimes N \xrightarrow{1 \otimes T \otimes 1} \Gamma \otimes \Gamma \otimes M \otimes N \xrightarrow{\mu \otimes 1 \otimes 1} \Gamma \otimes M \otimes N.$$

We call this the *diagonal coaction*, although perhaps a better name would be the "codiagonal" coaction. Note that one can define it using the push-forward construction described in the previous paragraph: $M \otimes N$ has the structure of a $\Gamma \otimes \Gamma$-comodule, and so inherits a Γ-comodule structure via the coalgebra map

$\mu \colon \Gamma \otimes \Gamma \to \Gamma$. We can also put the *left coaction* on $M \otimes N$, in which the structure map is
$$M \otimes N \xrightarrow{\psi_M \otimes 1} \Gamma \otimes M \otimes N.$$
Note that this completely ignores the comodule structure on N. We rarely use the left comodule structure on the tensor product; when we do, we denote the comodule by $M \overset{L}{\otimes} N$. If we want to explicitly distinguish the diagonal coaction from the left coaction, we write $M \overset{\Delta}{\otimes} N$ for the tensor product with the diagonal coaction.

We will use the following lemma once or twice. It is fairly standard; see [**Boa**, Theorem 5.7], or [**Mar83**, Proposition 12.4] for the module version.

LEMMA 0.3.2. *Let M be a Γ-comodule. Then $\Gamma \overset{\Delta}{\otimes} M$ with the diagonal coaction is naturally isomorphic, as a Γ-comodule, to $\Gamma \overset{L}{\otimes} M$ with the left coaction. In particular, $\Gamma \overset{\Delta}{\otimes} M$ is isomorphic to a direct sum of copies of Γ.*

PROOF. One can check that the following two composites are mutually inverse k-vector space maps:
$$\Gamma \overset{L}{\otimes} M \xrightarrow{1 \otimes \psi_M} \Gamma \otimes \Gamma \otimes M \xrightarrow{1 \otimes \chi \otimes 1} \Gamma \otimes \Gamma \otimes M \xrightarrow{\mu \otimes 1} \Gamma \otimes M,$$
$$\Gamma \otimes M \xrightarrow{1 \otimes \psi_M} \Gamma \otimes \Gamma \otimes M \xrightarrow{\mu \otimes 1} \Gamma \overset{L}{\otimes} M.$$
Furthermore, the second of these is easily seen to be a Γ-comodule map, and hence an isomorphism; hence the first composite is also a Γ-comodule isomorphism. $\square$

Starting in Chapter 2, we will work with comodules over A, the dual of the mod p Steenrod algebra. Since many readers may be more familiar with the study of modules over A^*, the Steenrod algebra, we include a brief discussion of the relationship between comodules over a Hopf algebra Γ and modules over its dual Γ^*.

LEMMA 0.3.3. *Let Γ be a coalgebra over k. Let Γ^* denote the graded dual of Γ; this is a k-algebra. Every Γ-comodule has a natural Γ^*-module structure, and if Γ is finite-dimensional, then the categories Γ-Comod and Γ^*-Mod are equivalent.*

PROOF. Let M be a Γ-comodule. Then we make it a Γ^*-module via the structure map
$$(0.3.4) \qquad \Gamma^* \otimes M \xrightarrow{1 \otimes \psi} \Gamma^* \otimes \Gamma \otimes M \xrightarrow{\text{ev} \otimes 1} k \otimes M = M.$$
If Γ is finite-dimensional, one can easily dualize this: given dual bases (γ_i) and (g_i) for Γ and Γ^*, then the dual of the evaluation map is
$$k \xrightarrow{\text{ev}^*} \Gamma \otimes \Gamma^*,$$
$$1 \longmapsto \sum_i \gamma_i \otimes \chi(g_i).$$
Composing with the module structure map ϕ makes M into a Γ-comodule:
$$M \xrightarrow{\text{ev}^* \otimes 1} \Gamma \otimes \Gamma^* \otimes M \xrightarrow{1 \otimes \phi} \Gamma \otimes M,$$
$$m \longmapsto \sum_j \gamma_j \otimes \phi(\chi(g_j) \otimes m).$$
We leave the details of the proof to the reader; see also [**HPS97**, Lemma 9.5.3]. $\square$

Regardless of whether Γ is finite-dimensional, the composite (0.3.4) defines a functor
$$J\colon \Gamma\text{-Comod} \to \Gamma^*\text{-Mod}.$$
Now, the functor J defines an equivalence between the category of Γ-comodules and the category of Γ^*-modules satisfying a finiteness condition called "tameness": a module M is *tame*, or *locally finite*, if every element $m \in M$ generates a finite submodule of M. The functor J has a right adjoint R, where RM is the largest tame submodule of M—see [**HPS97**, Lemma 9.5.3]. One can use R to convert operations on modules into similar operations for comodules. For example, if Γ^* is a Hopf algebra, given two Γ^*-modules M and N, then $\operatorname{Hom}_k(M, N)$ is naturally a Γ^*-module, via the "conjugation" action:
$$\Gamma \otimes \operatorname{Hom}_k(M, N) \to \operatorname{Hom}_k(M, N),$$
$$\gamma \otimes f(-) \longmapsto \sum \gamma' f(\chi(\gamma'') \cdot -),$$
where $\Delta(\gamma) = \sum \gamma' \otimes \gamma''$. If Γ is a Hopf algebra and M and N are Γ-comodules, or equivalently tame Γ^*-modules, then $\operatorname{Hom}_k(M, N)$ need not be tame. To fix this, one instead uses $R\operatorname{Hom}_k(M, N)$ as the internal Hom object in the comodule category. This is well-behaved; for instance, it is right adjoint to the tensor product. See [**HPS97**, Lemma 9.5.3] for more details.

REMARK 0.3.5. The above discussion gives some idea of the relative advantages and disadvantages of the two categories Γ-Comod and Γ^*-Mod. One has more flexibility when working with modules, whereas the finiteness restrictions on comodules may lead to some extra structure. In the module setting, for example, the "duality" functor $d\colon M \longmapsto \operatorname{Hom}_k(M, k)$ satisfies $ddM \cong M$ for any M which is of finite type. This is not true for comodules, in general; for example, if $\Gamma = A$ is the dual of the mod p Steenrod algebra, then $dA = R\operatorname{Hom}_{\mathbf{F}_p}(A, \mathbf{F}_p)$ is actually zero. On the other hand, "a comodule without generators is zero": if M is a Γ-comodule so that $k \,\square_\Gamma M = 0$, then $M = 0$. The corresponding statement fails for modules, as Margolis points out in [**Mar83**, Proposition 13.9]: when $\Gamma^* = A^*$, then $\mathbf{F}_p \otimes_{A^*} A = 0$.

0.4. Homological algebra

Now we discuss a little homological "coalgebra." [**Boa**] is a good reference for this material, and some of it may also be found in [**HPS97**, Section 9.5]. One can also dualize discussions of homological algebra for modules, as found in any number of places (such as [**CE56, Wei94, Ben91a**]).

Since we are working with comodules rather than modules, we work with the notions of cofree and injective comodules, which are dual to the notions of free and projective, respectively.

DEFINITION 0.4.1. Let Γ be a k-coalgebra. A Γ-comodule M is *injective* if the functor $\operatorname{Hom}_\Gamma(-, M)$ is exact. A comodule M is *projective* if $\operatorname{Hom}_\Gamma(M, -)$ is exact. The forgetful functor $U\colon \Gamma\text{-Comod} \to k\text{-Mod}$ has a right adjoint, C:
$$\operatorname{Hom}_k(UM, V) \cong \operatorname{Hom}_\Gamma(M, CV).$$
CV is called the *cofree* comodule on V; see Lemma 0.4.3 for a formula for C.

See [**Boa**] or [**HPS97**, Lemma 9.5.4] for the following results.

LEMMA 0.4.2. *A comodule is injective if and only if it is a summand of a cofree comodule.*

LEMMA 0.4.3. *For any vector space V, $CV \cong \Gamma \otimes V$ as k-vector spaces, with Γ-coaction $\Delta \otimes 1$. In particular (with $V = k$), Γ is injective as a Γ-comodule.*

As a consequence, every comodule map $M \to \Gamma$ is adjoint to a vector space map $M \to k$.

Injective comodules are more important to us than projective comodules, because of the following.

EXAMPLE 0.4.4. (a) On one hand, if Γ is a finite-dimensional commutative Hopf algebra, then injective comodules are the same as projective comodules: by a result of Larson and Sweedler [**LS69**, p. 85], finite-dimensional cocommutative Hopf algebras are self-injective as modules, so in the module setting, projective and injective are the same. Applying Lemma 0.3.3 yields the result for comodules over a commutative Hopf algebra.

(b) On the other hand, if $\Gamma = A$ is the dual of the mod p Steenrod algebra, then it seems likely that there are no nonzero projective comodules. As in the previous section, we can view any A-comodule as a module over the Steenrod algebra A^*, and in the category of A^*-modules, A is not projective—see [**Mar83**, Proposition 13.8]. This provides evidence that it is not projective as an A-comodule. Furthermore, we note that no finite-dimensional comodule has a projective comodule mapping onto it. To see this, we first observe that every finite-dimensional comodule maps surjectively to the trivial comodule $\mathbf{F}_p$, so it suffices to show that $\mathbf{F}_p$ has no projective comodule mapping onto it. For each integer $n \geq 1$, let M_n be the n-dimensional comodule which is spanned as a vector space by classes
$$\{y_0, y_1, \ldots, y_{n-1}\},$$
where the degree of y_i is $-|\xi_1|(1 + p + \cdots + p^{i-1})$. The comodule structure is given by
$$y_i \longmapsto 1 \otimes y_i + \sum_{j=i+1}^{n-1} \xi_1^{p^i + \cdots + p^{j-1}} \otimes y_j.$$

In other words, $y_i \longmapsto 1 \otimes y_i + \xi_1^{p^i} \otimes y_{i+1} + \text{(other terms)}$, where the "other terms" are of the form $a_j \otimes y_j$ where a_j is a non-primitive in A. For each n, there is a surjection $M_n \twoheadrightarrow \mathbf{F}_p$ which sends y_0 to 1 and every other y_i to 0. Given any projective comodule P and any map $f \colon P \to \mathbf{F}_p$, suppose there were a class $x \in P$ not in the kernel of f. For each n, consider the diagram

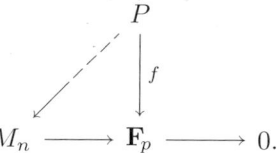

Since P is projective, there must be a lift $P \to M_n$. If $x \in P$ maps nontrivially to $\mathbf{F}_p$, then under the lift $P \to M_n$, x must map to a nonzero scalar multiple of y_0. Since y_0 has n terms in its diagonal, then x must

have at least n terms in its diagonal; since n is arbitrary, then x must have infinitely many terms in its diagonal. This can't happen, so every $x \in P$ must be in the kernel of $f \colon P \to \mathbf{F}_p$. In other words, if P is a projective comodule, then zero is the only comodule map $P \to \mathbf{F}_p$.

For any Γ-comodule M, Lemma 0.3.2 says that $\Gamma \otimes M$ is isomorphic to $\Gamma \otimes UM = CUM$, and hence is injective. The map

$$M = k \otimes M \xrightarrow{\eta \otimes 1} \Gamma \otimes M$$

which is adjoint to the identity on UM gives us the start of an injective resolution of M.

DEFINITION 0.4.5. Given a comodule M, a sequence of injective comodules $I_\bullet = (I_0 \to I_1 \to I_2 \to \cdots)$ together with a map $M \to I_0$ is an *injective resolution* of M if the sequence

$$0 \to M \to I_0 \to I_1 \to I_2 \to \cdots$$

is exact.

It is well-known that injective resolutions are unique up to cochain homotopy equivalence. We are interested in these resolutions because they are used to compute derived functors, such as Ext.

DEFINITION 0.4.6. Let Γ be a coalgebra over k, and let M and N be Γ-comodules. Then $\operatorname{Ext}^s_\Gamma(M,N)$ is the sth derived functor of $\operatorname{Hom}_\Gamma(M,N)$. To compute it, one takes an injective resolution $I_\bullet$ of N, and defines $\operatorname{Ext}^s_\Gamma(M,N)$ to be the sth cohomology group of the cochain complex $\operatorname{Hom}_\Gamma(M, I_\bullet)$.

Throughout this book, when we say Ext, we mean Ext in this setting: the category of comodules over a coalgebra.

Note that since Hom is graded, then Ext^s is graded for each s; in other words, Ext is bigraded. We write $\operatorname{Ext}^{s,t}$ for the degree t part of Ext^s, and we write Ext^{**} for $\bigoplus_{s,t} \operatorname{Ext}^{s,t}$.

The following two lemmas are standard, and easy, computations. We leave the details to the reader.

LEMMA 0.4.7. *Let Γ be a coalgebra over a field k. For any comodules M and N, the graded vector space $\operatorname{Ext}^0_\Gamma(M,N)$ is isomorphic to $\operatorname{Hom}^*_\Gamma(M,N)$.*

LEMMA 0.4.8. *Let Γ be a coalgebra over a field k. Then $\operatorname{Ext}^1_\Gamma(k,k)$ is isomorphic to $P\Gamma$, the space of primitives of Γ.*

REMARK 0.4.9. We have the following remarks, with which the reader is probably familiar.
   (a) We point out that for any comodule, there is a canonical injective resolution, known as the *cobar complex*. We will do a few computations with it in Appendix B.3; we refer the reader to [**Ada56**] and [**HMS74**] for details.
   (b) $\operatorname{Ext}^{**}_\Gamma(M,N)$ has important naturality properties: of course, it is functorial in M and N, contravariant in M and covariant in N. One can use the cobar complex to show that it is also natural in the coalgebra Γ, covariantly.
   (c) Finally, we note that $\operatorname{Ext}^{**}_\Gamma(k,k)$ is a bigraded k-algebra, commutative (in the graded sense) if Γ is a commutative Hopf algebra. The algebra structure is given by the "Yoneda product"; see [**Ben91a**, Section 2.6] for a description. One can use naturality, together with a Künneth isomorphism, to explain the commutativity: the product map $\mu \colon \Gamma \otimes \Gamma \to \Gamma$ is a

coalgebra map, so induces a map

$$\mu_*\colon \operatorname{Ext}^{**}_{\Gamma\otimes\Gamma}(k,k) \to \operatorname{Ext}^{**}_{\Gamma}(k,k).$$

Since we are tensoring over a field, then there is an isomorphism

$$\operatorname{Ext}^{**}_{\Gamma\otimes\Gamma}(k,k) \cong \operatorname{Ext}^{**}_{\Gamma}(k,k) \otimes \operatorname{Ext}^{**}_{\Gamma}(k,k).$$

One can check that the composite map $\mu_*\colon \operatorname{Ext}^{**}_{\Gamma}(k,k) \otimes \operatorname{Ext}^{**}_{\Gamma}(k,k) \to \operatorname{Ext}^{**}_{\Gamma}(k,k)$ agrees with the Yoneda product. Since the product μ is commutative, then so is μ_*.

We use the following result several times, so we state it here for the readers' convenience. This first appeared in [**HS98**], and is a generalization of results in [**Wil81**].

THEOREM 0.4.10 (Theorem 4.13 in [**HS98**]). *Suppose that $\Gamma \twoheadrightarrow \Lambda$ is a surjection of finite-dimensional graded connected commutative Hopf algebras over a field k of characteristic $p > 0$. For any $\lambda \in \operatorname{Ext}^{**}_{\Lambda}(k,k)$, there is a number m so that λ^{p^m} is in the image of the restriction map $\operatorname{Ext}^{**}_{\Gamma}(k,k) \to \operatorname{Ext}^{**}_{\Lambda}(k,k)$.*

In [**HS98**], this was stated for Ext of modules, not comodules; we have dualized it.

0.5. Two small examples

In this section, we discuss two small coalgebras, their comodules, and homological algebra over them. One can understand these coalgebras pretty thoroughly, and it is common practice to try to reduce the study of more complicated Hopf algebras and coalgebras to these simpler ones.

We work over a field k of characteristic $p > 0$.

NOTATION 0.5.1. Given a homogeneous element x in a graded vector space, recall that $|x|$ denotes its degree. Let $E[x]$ denote the coalgebra with basis $\{1, x\}$ with x primitive (Definition 0.2.4), where $|x|$ is odd if p is odd. Let $D[x]$ be the coalgebra with basis $\{x_0 = 1, x_1 = x, x_2, \ldots, x_{p-1}\}$, where $|x|$ is even if p is odd, where $|x_n| = n|x|$, and with coproduct given by this formula:

$$(0.5.2) \qquad \Delta(x_n) = \sum_{i=0}^{n} \binom{n}{i} x_i \otimes x_{n-i}.$$

These coalgebras usually arise in this book as Hopf algebras with algebra structure given by $E[x] \cong k[x]/(x^2)$ and $D[x] \cong k[x]/(x^p)$. Another algebra structure arises by viewing $D[x]$, for example, as being dual to the mod p group algebra of a cyclic group of order p. When we refer to "the Hopf algebra $E[x]$," we mean any Hopf algebra which is isomorphic as a coalgebra to $E[x]$; the same goes for $D[x]$.

Note that we can choose a different basis $\{y_0 = 1, y_1 = x, y_2, \ldots, y_{p-1}\}$ for $D[x]$ so that the coproduct formula is given by

$$(0.5.3) \qquad \Delta(y_n) = \sum_{i=0}^{n} y_i \otimes y_{n-i}.$$

(The dual situation is perhaps more familiar: given a k-algebra with basis $\{z_0 = 1, z_1, \ldots, z_{p-1}\}$ and product $z_i z_j = \binom{i+j}{i} z_{i+j}$ if $i + j < p$, then this is isomorphic to $k[z]/(z^p)$—just let $z = z_1$.)

Now we classify the finite-type comodules over these coalgebras.

PROPOSITION 0.5.4. *Suppose that k is a finite field and $m \geq 1$ is an integer. Let R be the k-algebra $k[x]/(x^m)$, graded with $|x| > 0$.*
  (a) *Every finite-type R-module may be written uniquely as a direct sum of indecomposable modules.*
  (b) *Up to suspension and isomorphism, the modules $k[x]/(x^n)$, with $1 \leq n \leq m$, are the only nonzero indecomposable R-modules.*

PROOF. Part (a) is a special case of [**Mar83**, Theorem 11.21]; note that this result is for algebras defined over a finite field.

Part (b): we classify all indecomposable $k[x]/(x^m)$-modules by induction on m. First we note that for any m, the algebra R is a *Poincaré algebra* (see [**Mar83**, p. 188]); hence every free R-module is both projective and injective.

When $m = 1$, then $R = k$; hence the result holds. Suppose that the result holds for modules over $k[x]/(x^{m-1})$, and let M be a nonzero indecomposable $k[x]/(x^m)$-module. If there is an element $y \in M$ so that $x^{m-1}y$ is nonzero, then the submodule generated by y is free, hence injective; hence it splits off of M as a summand. Since M is indecomposable, this submodule must be all of M; in other words, M is isomorphic to $k[x]/(x^m)$.

The other possibility is that $x^{m-1}y = 0$ for all $y \in M$. In this case, M is an indecomposable module over $k[x]/(x^{m-1})$. By induction, then, M is isomorphic to $k[x]/(x^n)$ for some n with $1 \leq n \leq m-1$. □

The coproduct (0.5.3) in $D[x]$ defines a comodule structure on the vector space $\mathrm{Span}(y_0, y_1, \ldots, y_{n-1})$ for each $n \leq p$. We write M_n for this n-dimensional comodule.

COROLLARY 0.5.5. (a) *Suppose that k is a finite field. Then every finite-type $E[x]$-comodule may be written uniquely as a direct sum of trivial comodules and cofree comodules.*
  (b) *Suppose that k is a finite field. Then every finite-type $D[x]$-comodule may be written uniquely as a direct sum of suspensions of the comodules M_n, $1 \leq n \leq p$.*

PROOF. By Lemma 0.3.3, the category of comodules over the coalgebra $E[x]$ (or $D[x]$) is equivalent to the category of modules over the dual algebra $k[x]/(x^2)$ (or $k[x]/(x^p)$), so Proposition 0.5.4 applies to the coalgebras of interest. □

Next we work out some homological algebra over $E[x]$ and $D[x]$. Both the methods and results will be useful later.

As in Section 0.1, we write s for the suspension functor on graded comodules. We note that there is a short exact sequence of $E[x]$-comodules

(0.5.6) $$0 \to k \to E[x] \to s^{|x|}k \to 0.$$

Repeatedly splicing this together with itself (and removing the k from the start) yields this sequence, which is an injective resolution of k:

(0.5.7) $$0 \to E[x] \to s^{|x|}E[x] \to s^{2|x|}E[x] \to \cdots \to s^{n|x|}E[x] \to \cdots.$$

This sequence is *periodic*: each map $s^{n|x|}E[x] \to s^{(n+1)|x|}E[x]$ sends x to 1 and sends 1 to 0. Note that one can extend this sequence to the left, as well, yielding an unbounded sequence which is exact:

$$\cdots \to s^{(n-1)|x|}E[x] \to s^{n|x|}E[x] \to s^{(n+1)|x|}E[x] \to \cdots.$$

Now we work with $D[x]$. As above, we let $M_n = \mathrm{Span}(y_0, y_1, \ldots, y_{n-1})$, with comodule structure given by formula (0.5.3). For each $n < p$, there are two relevant short exact sequences of $D[x]$-comodules:

(0.5.8)
$$0 \to M_n \to D[x] \to s^{n|x|} M_{p-n} \to 0,$$
$$0 \to M_{p-n} \to D[x] \to s^{(p-n)|x|} M_n \to 0.$$

Splicing these together alternately yields this injective resolution of M_n:

(0.5.9) $0 \to D[x] \to s^{n|x|} D[x] \to s^{p|x|} D[x] \to s^{(p+n)|x|} D[x] \to s^{2p|x|} D[x] \to \cdots.$

This sequence is periodic with period two. As with the $E[x]$ sequence, one can extend this to the left to get an unbounded long exact sequence.

One can use these injective resolutions to compute Ext with coefficients in any indecomposable comodule.

PROPOSITION 0.5.10.   (a) *As algebras, $\mathrm{Ext}^{**}_{E[x]}(k,k) \cong k[h]$, where the element $h \in \mathrm{Ext}^{1,|x|}_{E[x]}(k,k)$ corresponds to the short exact sequence (0.5.6).*

(b) *As algebras, $\mathrm{Ext}^{**}_{D[x]}(k,k) \cong \Lambda[h] \otimes k[b]$, where $h \in \mathrm{Ext}^{1,|x|}_{D[x]}(k,k)$ corresponds to the extension*
$$0 \to k \to M_2 \to s^{|x|} k \to 0,$$
and $b \in \mathrm{Ext}^{2,p|x|}_{D[x]}(k,k)$ corresponds to the composite of the two extensions given in (0.5.8):
$$0 \to k \to D[x] \to s^{|x|} D[x] \to s^{p|x|} k \to 0.$$

(c) *For any n with $1 \leq n \leq p-1$ and for any $i \geq 0$,*
$$\mathrm{Ext}^i_{D[x]}(k, M_n) \cong \begin{cases} s^{jp|x|} k & \text{if } i = 2j, \\ s^{(jp+n)|x|} k & \text{if } i = 2j+1. \end{cases}$$

This is a bigraded module over $\mathrm{Ext}^{**}_{D[x]}(k,k)$. Multiplication by b gives an isomorphism between $\mathrm{Ext}^i_{D[x]}(k, M_n)$ and $s^{-p|x|} \mathrm{Ext}^{i+2}_{D[x]}(k, M_n)$.

The algebra and module structures are not immediate from the resolutions, but they are not hard to compute, and are also standard.

CHAPTER 1

Stable homotopy over a Hopf algebra

In this chapter we discuss stable homotopy theory over a graded commutative Hopf algebra Γ over a field k; a major focus of study is $\operatorname{Ext}_\Gamma^{**}(k,k)$, where Ext denotes *comodule* Ext—derived functors of Hom in the category of left Γ-comodules. This material applies when Γ is the dual of a group algebra, the dual of an enveloping algebra, or the dual of the Steenrod algebra; in these cases, $\operatorname{Ext}_\Gamma^{**}(k,k)$ is the ordinary cohomology of the dual Γ^* with coefficients in k.

Our goals in this chapter are to establish some notation, make some basic definitions, and prove some general facts about the category Stable(Γ) of cochain complexes of injective Γ-comodules.

In more detail: we start in Section 1.1 by defining the setting for the rest of the book, the category Stable(Γ). We also set up some important notation; for instance, we explain the grading conventions on morphisms and (co)homology functors in Stable(Γ)—if X is an injective resolution of a left comodule M, then the (s,t)-homotopy group $\pi_{s,t}X$ is equal to $\operatorname{Ext}_\Gamma^{s,t}(k,M)$. In Section 1.2 we construct some particular ring objects in our category, one object HB for every quotient Hopf algebra B of Γ. To be precise, HB is an injective resolution of $\Gamma\,\square_B k$; this is a ring spectrum in Stable(Γ). So for instance, if we were working with $\Gamma = kG^*$, we would have one such object $H(kB^*)$ for every subgroup B of G, and the object $H(kB^*)$ would have homotopy groups $\pi_*H(kB^*) = H^*(B;k)$. In Subsection 1.2.1 we establish some notation for Hopf algebra extensions, and we prove one or two useful results about extensions with small kernel. For example, given an extension of Hopf algebras of the form

$$E[x] \to B \to C,$$

the associated extension spectral sequence has only one possible differential; if B is a quotient Hopf algebra of Γ, then this manifests itself in the category Stable(Γ) as a cofibration sequence $HB \to HB \to HC$. In Section 1.3 we set up cellular towers and Postnikov towers in Stable(Γ), and we prove a Hurewicz theorem and a few useful lemmas. In Section 1.4 we discuss generalized Adams spectral sequences in Stable(Γ), focusing on the spectral sequence based on the homology theory associated to the ring spectrum HB, for B a "conormal" quotient of Γ. This turns out to be the same, up to a regrading, as the Lyndon-Hochschild-Serre spectral sequence associated to the Hopf algebra extension

$$\Gamma\,\square_B k \to \Gamma \to B.$$

In Section 1.5 we define Bousfield classes and Brown-Comenetz duality, and we recall some results of Ravenel's relating the two.

In Section 1.6 we apply some of this work to the study of stable homotopy over a group algebra. We point out, for example, that a corollary of work of Benson,

Carlson, and Rickard is a classification of the Bousfield lattice in Stable(kG^*), for G a p-group and k a field of characteristic p.

1.1. The category Stable(Γ)

In this section, we define the category in which we work, and we introduce notation which we will use for the rest of this book.

Let Γ be a graded commutative Hopf algebra over a field k; we work in the category Stable(Γ). The objects of this category are unbounded **Z**-graded cochain complexes of injective left Γ-comodules; we call these objects *spectra*. The morphisms are the cochain homotopy classes of bigraded maps; to be more precise, we first describe the bigrading on the category. One can shift each object in both the "cohomological direction" (changing the cochain complex grading) and the "internal direction" (since the cochain complexes consist of graded comodules). More precisely, given an object

$$X = (\cdots \to I_{-1} \to I_0 \to I_1 \to \cdots),$$

we let $\Sigma^{i,j}X$ be the cochain complex which is $s^j I_{n-i}$ in homological degree n, with the apparent differential; here s denotes the internal suspension functor on the category Γ-Comod, as described in Section 0.1. $\Sigma^{i,j}$ is called the (i,j)-*suspension functor*.

For objects X and Y of Stable(Γ), we let $[X,Y]_{0,0}$ denote the cochain homotopy classes of graded maps of bidegree $(0,0)$ from X to Y, and we let $[X,Y]_{i,j} = [\Sigma^{i,j}X, Y]_{0,0}$. The set of morphisms in Stable(Γ) from X to Y is the set

$$[X,Y]_{**} = \bigoplus_{i,j} [X,Y]_{i,j}$$

of cochain homotopy classes of all graded maps. If we want to indicate the Hopf algebra Γ, we write the morphisms as $[X,Y]_{**}^{\Gamma}$. As with maps of graded vector spaces, we often omit the degree of any given map, so the notation $X \to Y$ indicates an element in $[X,Y]_{**}$, not necessarily an element in $[X,Y]_{0,0}$. We have chosen this grading so that it matches up well with Ext: see Remark 1.1.2.

The category Stable(Γ) is a *stable homotopy category* in the sense of [**HPS97**]; hence one can perform many standard stable homotopy theoretic constructions in it. (See [**HPS97**, Theorem 9.5.1] for a verification that Stable(Γ) is a stable homotopy category.) For instance, rather than having exact sequences, one has "exact triangles," also known as "cofibrations" or "cofiber sequences." We freely use other language and results from [**HPS97**], often without explicit citation. Stable(Γ) is (weakly) *generated* by the injective resolutions of the simple comodules—that is, if X is an object so that $[J,X]_{**} = 0$ whenever J is an injective resolution of a simple comodule, then X is a contractible cochain complex. If the trivial comodule k and its suspensions are the only simples (say, if Γ is connected, or if Γ is the dual of the mod p group algebra of a p-group), then we say that Stable(Γ) is *monogenic*, at least in the graded sense. In this case, the stable homotopy constructions are even more familiar.

Let $\mathscr{S}$ denote the set of simple Γ-comodules. If the set

$$\bigcup_{T,T' \in \mathscr{S}} \operatorname{Ext}_{\Gamma}^{**}(T,T')$$

is countable, then Stable(Γ) is a *Brown category*, so that homology functors are representable. This is the case when $\Gamma = A$, the dual of the Steenrod algebra.

REMARK 1.1.1. (a) As in any stable homotopy category, honest limits and colimits usually do not exist in Stable(Γ); rather, one has to work with homotopy limits and colimits. Hence the notation lim and colim stand for the homotopy versions. See [**HPS97**, Section 2.2] for information about limits and colimits in stable homotopy categories.

(b) Hovey [**Hov99**] has constructed a closed model category so that Stable(Γ) is equivalent to the associated homotopy category. A summary of the relevant results is given in Appendix A below.

(c) In the case when $\Gamma = A$ is the dual of the mod p Steenrod algebra, Mahowald and Sadofsky studied the category Stable(Γ) in their paper [**MS95**].

(d) If Γ is finite-dimensional, then one could just as well work with the category of cochain complexes of injective Γ^*-modules, because in the finite-dimensional case, the categories of Γ-comodules and Γ^*-modules are equivalent, with injective comodules corresponding to injective modules. When Γ^* is not finite-dimensional, in particular when $\Gamma^* = A^*$ is the Steenrod algebra, there are technical problems with the category of cochain complexes of injective A^*-modules. For example, there are no maps from $\mathbf{F}_p$ to A^*, so the "homotopy" of the injective module A^* would be zero; therefore in the module setting, we would not have the implication $\pi_{**}X = 0 \Rightarrow X = 0$. See the discussion at the end of Section 0.3 for more information.

(e) We also note that, regardless of the dimension of Γ, the category Stable(Γ) is rather different from the derived category of Γ-comodules, because homology isomorphisms are not necessarily invertible in Stable(Γ). For instance, if $\Gamma = E[x]$ with x primitive, then the periodic cochain complex

$$\cdots \to \Gamma \to \Gamma \to \Gamma \to \cdots,$$

in which each map sends x to 1 and 1 to 0, has no homology, and hence is zero in the derived category. On the other hand, it is non-contractible in Stable(Γ); if we write $\operatorname{Ext}_\Gamma^{**}(k,k) = k[v]$, then this complex is a ring spectrum with homotopy groups (as defined below) equal to $k[v, v^{-1}]$.

The category Stable(Γ) has arbitrary *coproducts*; we use the symbol $\vee$ to denote the coproduct. We sometimes use the word *wedge* as a synonym for coproduct.

For objects X and Y of Stable(Γ), we write $X \wedge Y$ for $X \otimes_k Y$, and we call this the *smash product* of X and Y. This operation is commutative, associative, and unital: if S is an injective resolution of the trivial comodule k, then S is the unit of the smash product. We call S the *sphere* spectrum. We write $S^{i,j}$ for $\Sigma^{i,j}S$, so that $\Sigma^{i,j}X = S^{i,j} \wedge X$ for any X.

REMARK 1.1.2. The grading is the usual Ext grading: if X and Y are injective resolutions of comodules M and N, respectively, then $[X,Y]_{i,j} = \operatorname{Ext}_\Gamma^{i,j}(M,N)$. Hence if Γ is concentrated in degree 0 (e.g., if $\Gamma = (kG)^*$), then one may as well work with Γ-comodules concentrated in degree 0, in which case $[-,-]_{ij} = 0$ if $j \neq 0$, and $[-,-]_{i0} = \operatorname{Ext}_\Gamma^i(-,-)$. We will follow the (somewhat odd) tradition in homotopy theory of drawing pictures of Ext (and hence of $[-,-]$) using the Adams spectral sequence grading: $\operatorname{Ext}^{s,t}$ is drawn with s on the vertical axis and $t-s$ on the horizontal axis.

Unfortunately, because of the form of long exact sequences in Ext, cofiber sequences look like this:

$$\cdots \to \Sigma^{1,0}Z \to X \to Y \to Z \to \Sigma^{-1,0}X \to \cdots.$$

So one needs to take a little care when translating proofs from ordinary homotopy theory to this setting. Given a spectrum X and integers i and j, we define the *homology functor* associated to X, X_{ij}, by

$$X_{ij}\colon \mathsf{Stable}(\Gamma) \to \mathsf{Ab},$$
$$Y \longmapsto [S, X \wedge Y]_{i,j},$$

and we define the *cohomology functor* associated to X, X^{ij}, by

$$X^{ij}\colon \mathsf{Stable}(\Gamma)^{\mathrm{op}} \to \mathsf{Ab},$$
$$Y \longmapsto [Y, X]_{-i,-j}.$$

When $X = S$, we have a special notation for X_{ij}: we define the (i,j)-*homotopy group* of Y to be $\pi_{ij}Y = S_{ij}Y = [S, Y]_{ij}$. As with morphisms, we write $\pi_{**}(-)$ for $\bigoplus_{i,j} \pi_{ij}(-)$ (and similarly for other homology and cohomology functors). Also, given a spectrum X, we write X_{ij} for $\pi_{ij}X = X_{ij}S$, and X_{**} for $\bigoplus_{i,j} X_{ij}$. Some people might refer to $X_{**} = \pi_{**}X$ as the "hypercohomology" of the cochain complex X, but we will try to restrain ourselves. Note that if $X \to Y \to Z$ is a cofiber sequence, then we have a long exact sequence

$$\cdots \to \pi_{i-1,j}Z \to \pi_{i,j}X \to \pi_{i,j}Y \to \pi_{i,j}Z \to \pi_{i+1,j}X \to \cdots$$

(and similarly for other homology functors).

We say that an object R in $\mathsf{Stable}(\Gamma)$ is a *ring spectrum* if there is a multiplication map $\mu\colon R \wedge R \to R$ and a unit map $\eta\colon S \to R$, making the appropriate diagrams commute.

We often abuse notation and let $S^0 = S^{0,0} = S$. One of our main goals is to get as much information as possible about $\pi_{**}S^0 = \mathrm{Ext}_\Gamma^{**}(k,k)$.

1.2. The functor H

We assume that Γ is a graded commutative Hopf algebra over a field k, and we work in the category $\mathsf{Stable}(\Gamma)$. The quotient coalgebras and Hopf algebras of Γ carry useful information; in this section we construct a spectrum HB in $\mathsf{Stable}(\Gamma)$ for each quotient coalgebra B of Γ, and we study the properties of the functor H.

Recall that if B is a quotient coalgebra of Γ and if M is a B-comodule, then the *cotensor product* $\Gamma \,\square_B M$ is defined to be the equalizer of the two maps

$$\Gamma \otimes M \xrightarrow{1_\Gamma \otimes \psi_M} \Gamma \otimes B \otimes M,$$
$$\Gamma \otimes M \xrightarrow{\psi_\Gamma \otimes 1_M} \Gamma \otimes B \otimes M.$$

Here ψ_Γ is the right B-comodule structure map on Γ, and ψ_M is the left B-comodule structure map on M. (The tensor products are over k, as usual.)

Alternatively, taking the cotensor product with Γ is right adjoint to restriction; see Lemma 1.2.3 below.

DEFINITION 1.2.1. We define a covariant functor H from quotient coalgebras of Γ to spectra by defining HB to be an injective resolution of $\Gamma \,\square_B k$.

For example, $Hk = \Gamma$, so that $Hk_{**}(X)$ is the homology of the cochain complex X; also, $H\Gamma = S^0$. H provides a useful source of (co)homology functors on Stable(Γ). The general philosophy is that if one has a quotient B of Γ, rather than studying B by working in Stable(B), one studies B by looking at HB in the category Stable(Γ). This is borne out by Corollary 1.2.7, as well as the other results in this section.

In order to study the objects HB, we need to examine the cotensor product. We start with its adjointness property.

Given a quotient coalgebra B of Γ, we write $[-,-]^B$ for the set of cochain homotopy classes of B-comodule maps. We abuse notation and let Stable(B) denote the category with objects cochain complexes of injective B-comodules, and morphisms $[-,-]^B$. (This is an abuse of notation because the category Stable(B) need not be a stable homotopy category—it won't have a smash product unless B is a quotient Hopf algebra.)

DEFINITION 1.2.2. Fix a quotient coalgebra B of Γ. Given a Γ-comodule M, we let $M\!\downarrow_B$ denote its *restriction* to B; this is the B-comodule with structure map $M \to \Gamma \otimes M \to B \otimes M$. We occasionally write $\mathrm{res}_{\Gamma,B}$ for restriction. We use the same notations for related restriction functors, such as the one from Stable(Γ) to Stable(B).

Recall from [**MM65**] and [**Rad77**] that if B happens to be a quotient Hopf algebra of Γ, then $\Gamma\!\downarrow_B$ is injective as a right (and a left) B-comodule. When $\Gamma\!\downarrow_B$ is injective as a right B-comodule, the functor $\Gamma \square_B -\colon B\text{-Comod} \to \Gamma\text{-Comod}$ is exact and takes injectives to injectives; hence it induces a functor $\Gamma \square_B -\colon \mathsf{Stable}(B) \to \mathsf{Stable}(\Gamma)$, defined by applying $\Gamma \square_B -$ dimensionwise. Regardless of the context, we call $\Gamma \square_B -$ the *induction* functor, and we occasionally write $\mathrm{ind}_{B,\Gamma}$ instead of $\Gamma \square_B -$.

LEMMA 1.2.3. *The functors $-\!\downarrow_B$ and $\Gamma \square_B -$ are adjoint:*

(a) *Let B be a quotient coalgebra of Γ. Given a Γ-comodule M and a B-comodule N, there is an isomorphism*

$$\mathrm{Hom}_B(M\!\downarrow_B, N) \cong \mathrm{Hom}_\Gamma(M, \Gamma \square_B N),$$

natural in M and N.

(b) *Let B be a quotient coalgebra of Γ over which $\Gamma\!\downarrow_B$ is injective as a right B-comodule. Then given objects $X \in \mathsf{Stable}(\Gamma)$ and $Y \in \mathsf{Stable}(B)$, the isomorphism of part (a) induces a natural isomorphism*

$$[X\!\downarrow_B, Y]^B \cong [X, \Gamma \square_B Y]^\Gamma.$$

Part (a) is dual to the classical statement for modules that induction $M \mapsto \Gamma \otimes_B M$ is left adjoint to restriction; see [**Ben91a**, Proposition 2.8.3], for instance.

PROOF. Part (a): Define

$$\alpha\colon \mathrm{Hom}_B(M\!\downarrow_B, N) \to \mathrm{Hom}_\Gamma(M, \Gamma \square_B N)$$

by $\alpha(f) = (1 \otimes f) \circ \psi_M$, where ψ_M is the Γ-coaction map on M. One can verify that $\alpha(f)$ is a Γ-comodule map for any $f \in \mathrm{Hom}_B(M\!\downarrow_B, N)$. Define

$$\beta\colon \mathrm{Hom}_\Gamma(M, \Gamma \square_B N) \to \mathrm{Hom}_B(M\!\downarrow_B, N)$$

by $\beta(g) = (\varepsilon \otimes 1) \circ g$; one can verify that $\beta(g)$ is a map of B-comodules. One can also check that α and β are inverses. We leave the details to the reader; the verifications are dual to the corresponding verifications for modules.

Part (b): One needs to check two things: that the maps α and β of part (a) induce well-defined maps on sets of cochain maps, and that they induce well-defined maps on cochain homotopy classes of maps. These are both straightforward verifications, which we leave to the reader. □

EXAMPLE 1.2.4. If $\Gamma\!\downarrow_B$ is injective over B, M is a Γ-comodule and N is a B-comodule, then we have
$$\operatorname{Ext}_B^{**}(M\!\downarrow_B, N) \cong \operatorname{Ext}_\Gamma^{**}(M, \Gamma \square_B N).$$
This result is often called a "change-of-rings isomorphism" or "Shapiro's lemma"—see [**Rav86**, Theorem A1.3.12] and [**Ben91a**, Corollary 2.8.4], for instance. It follows from part (b) of the lemma by letting X be an injective resolution of M and Y an injective resolution of N.

LEMMA 1.2.5. *Suppose that B is a quotient coalgebra of Γ so that $\Gamma\!\downarrow_B$ is injective as a right B-comodule. If M is a Γ-comodule, then there is an isomorphism of Γ-comodules*
$$\Gamma \square_B (M\!\downarrow_B) \cong (\Gamma \square_B k) \otimes M.$$

This result is given for modules in [**HS98**, p. 23]—it is called the "shearing isomorphism" there.

PROOF. Note that $\Gamma \square_B (M\!\downarrow_B)$ is a Γ-subcomodule of $\Gamma \overset{L}{\otimes} M$ (where $\overset{L}{\otimes}$ indicates the tensor product with the left Γ-coaction), and $(\Gamma \square_B k) \otimes M$ is a Γ-subcomodule of $\Gamma \otimes M$ (with the usual, diagonal, coaction). One only needs to check that the isomorphism
$$\Gamma \overset{L}{\otimes} M \cong \Gamma \otimes M$$
of Lemma 0.3.2 carries these subcomodules to each other. We leave the details to the reader. □

Applied dimensionwise to a cochain complex X in $\mathsf{Stable}(\Gamma)$, we have
$$\Gamma \square_B (X\!\downarrow_B) \cong (\Gamma \square_B k) \otimes X.$$
Since HB is an injective resolution of $\Gamma \square_B k$, we have the following.

COROLLARY 1.2.6. *Suppose that B is a quotient coalgebra of Γ so that $\Gamma\!\downarrow_B$ is injective as a right B-comodule. For any object $X \in \mathsf{Stable}(\Gamma)$, there is a cochain homotopy equivalence $\Gamma \square_B (X\!\downarrow_B) \simeq HB \wedge X$.*

Combining this with Lemma 1.2.3 yields the following.

COROLLARY 1.2.7. *Suppose that B is a quotient coalgebra of Γ so that $\Gamma\!\downarrow_B$ is injective as a right B-comodule. Given objects X and Y of $\mathsf{Stable}(\Gamma)$, we have $[X, HB \wedge Y]_{**} \cong [X\!\downarrow_B, Y\!\downarrow_B]_{**}^B$. In particular, HB_{**} is an algebra and $HB_{**}Y$ is a right module over it.*

In summary, we have the following. Note that the result is stronger for quotient Hopf algebras than for quotient coalgebras.

PROPOSITION 1.2.8. *For any quotient coalgebra B of Γ over which $\Gamma \downarrow_B$ is injective as a right B-comodule, there is an isomorphism $HB_{ij} \cong \mathrm{Ext}_B^{ij}(k,k)$. If X is an injective resolution of a Γ-comodule M, then $HB_{**}X \cong \mathrm{Ext}_B^{**}(k, M\downarrow_B)$. Furthermore, we have the following.*

(a) *If B is a quotient Hopf algebra of Γ, then HB is a commutative associative ring spectrum.*
(b) *If B is a quotient coalgebra of Γ over which $\Gamma \downarrow_B$ is injective as a right comodule, then HB has many of the properties of a ring spectrum:*
   (i) *HB_{**} is a k-algebra, and for any spectrum Y, $HB_{**}Y$ is a right module over HB_{**}.*
   (ii) *There is a "unit map" $S^0 \to HB$ which induces an algebra map $\pi_{**}S^0 \to HB_{**}$.*
   (iii) *More generally, if $B \twoheadrightarrow C$ are quotient coalgebras of Γ over which Γ is injective, then the induced map $HB_{**} \to HC_{**}$ is an algebra map.*

PROOF. That $HB_{**} \cong \mathrm{Ext}_B^{**}(k,k)$ follows from Lemma 1.2.3.

Part (a) is clear—since Γ is commutative and associative as an algebra, then so is $\Gamma \square_B k$, and the commutative associative product on $\Gamma \square_B k$ induces one on HB. The quotient map $\Gamma \twoheadrightarrow B$ induces the unit map $S^0 = H\Gamma \to HB$, and hence a Hurewicz map $\pi_{**}(X) \to HB_{**}(X)$. This Hurewicz map is the same as the *restriction map* $\mathrm{res}_{\Gamma,B}$: $\mathrm{Ext}_\Gamma^{**}(k,k) \to \mathrm{Ext}_B^{**}(k,k)$.

For part (b), part (i) is part of Corollary 1.2.7. The unit map is induced from the quotient $\Gamma \twoheadrightarrow B$, as in (a), and the induced map in (iii) is just the restriction map. $\square$

EXAMPLE 1.2.9. If B is a quotient coalgebra of Γ, not a quotient Hopf algebra, then HB need not be a ring spectrum. For example, suppose that k has characteristic p, and let $\Gamma = k[x]$ with x primitive. Then $B = k[x^p]$ is a quotient coalgebra of Γ, and HB is an injective resolution of $M = k[x]/(x^p)$. A multiplication on HB would induce one on M (by taking homology), and it is easy to see that M is not a Γ-comodule algebra: the comodule structure on M is given by $x \mapsto 1 \otimes x + x \otimes 1 \in \Gamma \otimes M$, so if this coaction were multiplicative, we would have
$$0 = x^p \longmapsto x^p \otimes 1 \neq 0.$$
Hence HB is not a ring spectrum.

The example Hk, which represents the homology of cochain complexes, is particularly important.

COROLLARY 1.2.10. *The spectrum Hk is a field spectrum: it is a commutative ring spectrum, and any Hk-module spectrum is a wedge of suspensions of Hk. In particular, for any X, $Hk \wedge X$ is a wedge of suspensions of Hk.*

PROOF. By Proposition 1.2.8(a), $Hk = \Gamma$ is a commutative ring spectrum. By Lemma 0.3.2, for any Γ-comodule M, $\Gamma \otimes M$ is a direct sum of copies of Γ; one can check (by [**Boa**, Theorem 5.7], for instance) that given a comodule map $M \to N$, the induced map $\Gamma \otimes M \to \Gamma \otimes N$ sends each summand of Γ either isomorphically to a summand, or to zero. So if X is any cochain complex of Γ-comodules, then $Hk \wedge X = \Gamma \otimes X$ splits into a direct sum of cochain complexes of the forms
$$0 \to \Gamma \xrightarrow{=} \Gamma \to 0$$

and
$$0 \to \Gamma \to 0.$$

Hence we have the following.

COROLLARY 1.2.11. *Hk_{**} satisfies a Künneth isomorphism.*

1.2.1. Remarks on Hopf algebra extensions. Many results about group cohomology (and indeed about Hopf algebra cohomology in general) are proved using the spectral sequence associated to an extension. If one has a group extension in which the quotient is cyclic of prime order, the associated spectral sequence is particularly tractable, and hence quite useful. In this subsection we remind the reader of standard notation related to Hopf algebra extensions, and then we focus on extensions with small kernel. We observe that in this case, the spectral sequences degenerate to cofibrations. In Section 1.4 we discuss the spectral sequence associated to a general Hopf algebra extension.

DEFINITION 1.2.12. (a) Suppose that $C \hookrightarrow \Gamma$ is an inclusion of augmented algebras over a field k, and let IC be the augmentation ideal of C—the kernel of $C \to k$. We say that C is *normal* in Γ if the left ideal of Γ generated by C (i.e., $\Gamma \cdot IC$) is equal to the right ideal generated by C (i.e., $IC \cdot \Gamma$). If C is normal in Γ, then we let $\Gamma//C = \Gamma \otimes_C k = k \otimes_C \Gamma$. In this case,
$$C \to \Gamma \to \Gamma//C$$
is an extension of augmented k-algebras.

(b) Dually, suppose that $\Gamma \to B$ is a surjective map of coaugmented coalgebras over k, and let JB denote the coaugmentation coideal of B—the cokernel of $k \to B$. We say that B is a *conormal* quotient of Γ if $\Gamma \square_B k = k \square_B \Gamma$ as sub-vector spaces of Γ. If B is a conormal quotient of Γ, then
$$\Gamma \square_B k \to \Gamma \to B$$
is an extension of coaugmented k-coalgebras.

As far as this definition goes, for us Γ will usually be a Hopf algebra with commutative multiplication, so that every subalgebra of Γ will be normal. We will be more interested in quotients of Γ; if $\Gamma \to B$ is a surjective map of commutative Hopf algebras over k, then $\Gamma \square_B k$ is the algebra kernel. So B is conormal if the algebra kernel is a subcoalgebra of Γ and is closed under conjugation χ.

REMARK 1.2.13. Suppose that Γ is a commutative Hopf algebra and B is a conormal quotient coalgebra of Γ over which Γ is injective. For any Γ-comodule M, there is a coaction of $\Gamma \square_B k$ on $\operatorname{Ext}_B^{**}(k, M\downarrow_B)$, which may be constructed as follows. Let $I_\bullet$ be a Γ-injective resolution of M; then this Ext group is the cohomology of the cochain complex with terms $\operatorname{Hom}_B(k, I_n)$. For each n, $\operatorname{Hom}_B(k, I_n)$ is a $\Gamma \square_B k$-comodule, via the Γ-coaction on I_n, and the boundary maps in the cochain complex are $\Gamma \square_B k$-comodule maps. Hence $\operatorname{Ext}_B^n(k, M\downarrow_B)$ is a $\Gamma \square_B k$-comodule for each integer n. More generally, the same argument produces a $\Gamma \square_B k$-coaction on $HB_{n*}X$ for any object X in Stable(Γ), for each integer n.

For the remainder of this section, we assume that the ground field k has characteristic $p > 0$. We use the results of Section 0.5 here; in particular, see Notation 0.5.1 for the coalgebras $E[x]$ and $D[x]$.

NOTATION 1.2.14. Recall from Lemma 0.4.8 that if B is a Hopf algebra, then $\operatorname{Ext}_B^{1,*}(k,k) = HB_{1,*}$ is isomorphic to the vector space of primitives of B. If $y \in B$ is primitive, we let $[y]$ denote the associated element of $HB_{1,*}$. Recall further that if p is odd, then there is a Steenrod operation
$$\beta\widetilde{\mathscr{P}^0}\colon \operatorname{Ext}_B^{s,t}(k,k) \to \operatorname{Ext}_B^{s+1,pt}(k,k).$$
See [**May70**, **Wil81**], as well as Appendix B.1. Given $[y] \in HB_{1,*}$, the element $\beta\widetilde{\mathscr{P}^0}[y]$ is also the p-fold Massey product of $[y]$ with itself, as mentioned in [**May70**, Remarks 11.11].

As mentioned above, when studying group cohomology, one often uses the spectral sequence for a group extension of the form
$$1 \to H \to G \to \mathbf{Z}/p \to 1.$$
In our "dual" setting, we consider Hopf algebra extensions with small kernel. These give us cofibrations which we can use in place of spectral sequences.

LEMMA 1.2.15. *Fix a graded commutative Hopf algebra* Γ *over a field* k *of characteristic* $p > 0$.

(a) *Suppose that there is a Hopf algebra extension of the form*
$$E[x] \to B \to C,$$
where B is a quotient Hopf algebra of Γ. Then x is primitive in B, so there is a nonzero element $h = [x]$ in $\operatorname{Ext}_B^{1,|x|}(k,k) = HB_{1,|x|}$. We also let $h\colon \Sigma^{1,|x|}HB \to HB$ denote the corresponding self-map of HB—i.e., the composite
$$\Sigma^{1,|x|}HB = S^{1,|x|} \wedge HB \xrightarrow{h \wedge 1} HB \wedge HB \xrightarrow{\mu} HB,$$
where μ is the multiplication map. Then there is a cofiber sequence
$$\Sigma^{1,|x|}HB \xrightarrow{h} HB \to HC \to \Sigma^{0,|x|}HB.$$

(b) *Suppose that p is odd and there is a Hopf algebra extension of the form*
$$D[x] \to B \to C,$$
where B is a quotient Hopf algebra of Γ. Then x is primitive in B; we let $b = \beta\widetilde{\mathscr{P}^0}[x]$ in $\operatorname{Ext}_B^{2,p|x|}(k,k) = HB_{2,p|x|}$. We also let $b\colon \Sigma^{2,p|x|}HB \to HB$ denote the corresponding self-map of HB. Then there is a cofiber sequence
$$\Sigma^{2,p|x|}HB \xrightarrow{b} HB \to \widetilde{HC} \to \Sigma^{1,p|x|}HB,$$
where $\widetilde{HC}$ is defined by a cofibration
$$\Sigma^{1,|x|}HC \to \widetilde{HC} \to HC \to \Sigma^{0,|x|}HC.$$

PROOF. Part (a): It is clear that x is primitive, so we only need to discuss the putative cofibration. Equation (0.5.6) gives an exact sequence of $E[x]$-comodules:
$$0 \to k \to E[x] \to s^{|x|}k \to 0.$$
We apply the (exact) functor $\Gamma \square_B -$ to this, noting that $E[x] = B \square_C k$:
$$0 \to \Gamma \square_B k \to \Gamma \square_C k \to s^{|x|}\Gamma \square_B k \to 0.$$
Taking injective resolutions gives the desired cofibration.

Part (b): We apply the functor $\Gamma\square_B-$ to the two short exact sequences of $D[x]$-comodules in (0.5.8) and take injective resolutions; writing $\widetilde{HB}$ for an injective resolution of the comodule $\Gamma\square_B(k[x]/(x^{p-1}))$, we have the following cofiber sequences:

$$\Sigma^{1,|x|}\widetilde{HB} \xrightarrow{\eta} HB \to HC \to \Sigma^{0,|x|}\widetilde{HB},$$

$$\Sigma^{1,(p-1)|x|}HB \xrightarrow{\eta'} \widetilde{HB} \to HC \to \Sigma^{0,(p-1)|x|}HB.$$

It is standard (e.g., see [**Ben91b**, pp. 137–8]) that the map b is the composite $\eta \circ \eta'$, and the 3×3 lemma (or the octahedral axiom—see [**HPS97**, A.1.1–A.1.2]) allows us to identify the cofiber of $\eta \circ \eta'$ in terms of the cofibers of η and η'. □

1.3. Some classical homotopy theory

We assume that Γ is a graded commutative Hopf algebra over a field k, and we work in the category $\mathsf{Stable}(\Gamma)$. For this section, we assume that $\mathsf{Stable}(\Gamma)$ is monogenic—i.e., the trivial comodule k and its suspensions are the only simple comodules. We also assume that Γ is non-negatively graded: $\Gamma_n = 0$ if $n < 0$.

Because $\pi_{ij}S^0 = \mathrm{Ext}_\Gamma^{i,j}(k,k)$, then $\pi_{**}S^0$ is concentrated in the first quadrant. More precisely, $\pi_{ij}S^0 = 0$ if $j < 0$ and (unless $i = j = 0$) if $i \leq 0$; furthermore, $\pi_{00}S^0 = k$. Since $\pi_{**}S^0$ is "connected" in this sense, we can construct cellular towers and Postnikov towers, as in the usual stable homotopy category (and indeed in any *connective* stable homotopy category—see [**HPS97**, Section 7]). We also have a Hurewicz theorem. Since we are working in a bigraded setting rather than the singly graded setting of [**HPS97**, Section 7], we state the relevant results. See [**Mar83**, Chapter 3] for the proofs in the ordinary stable homotopy category; the reader can modify them for this setting fairly easily.

REMARK 1.3.1. A priori, since the boundary homomorphism in π_{**} raises degrees rather than lowers them, there could be difficulties in translating the proofs in [**Mar83**] to the category $\mathsf{Stable}(\Gamma)$. On the other hand, we are "working over a field," meaning that $\pi_{00}S^0$ is a field k. This means that every homotopy group is a k-module; hence, given any object X and any bidegree (i,j), one can find a map $\bigvee S^{i,j} \to X$ which induces not just an epimorphism on π_{ij}, but an isomorphism. Indeed, because of the connectivity properties of $\pi_{**}S^0$, for any numbers $m \leq 0$ and n, there is a map from a wedge of spheres to X inducing an isomorphism on π_{ij} for all i and j with $j = mi + n$. This feature compensates for any possible problems with the boundary homomorphism.

DEFINITION 1.3.2. Given a spectrum X and integers k and m, then the diagram

$$0 = X_k \to X_{k+1} \to X_{k+2} \to \cdots$$

is a *strong cellular tower* of slope m for X if
   (a) X is the sequential colimit of the X_n,
   (b) the cofiber of each map $X_n \to X_{n+1}$ is a wedge of spheres $S^{i,j}$ with $j = mi + n$.

In contrast, a *cellular tower* is a similar diagram in which the cofibers are wedges of spheres with no restrictions on their dimensions. According to [**HPS97**, Proposition 2.3.1], every object in $\mathsf{Stable}(\Gamma)$ has a cellular tower. Because of the connectivity properties of $\pi_{ij}S^0$, we have the following theorem, guaranteeing existence of strong cellular towers with and Postnikov towers.

THEOREM 1.3.3. *Let X be a spectrum and m a non-positive integer.*

(a) *Suppose that there is an integer k so that $\pi_{ij}X = 0$ whenever $j \leq mi + k$. Then X has a strong cellular tower of slope m:*
$$0 = X_k \to X_{k+1} \to X_{k+2} \to \cdots.$$

(b) *For any integer n, there is a cofiber sequence*
$$X[n,\infty]_m \to X \to X[-\infty, n-1]_m,$$
where
  (i) $\pi_{ij}(X[n,\infty]_m) = 0$ *if* $j < mi + n$,
  (ii) $\pi_{ij}(X[-\infty, n-1]_m) = 0$ *if* $j \geq mi + n$.

(c) *If for some integer n we have spectra Y and Z with $Y = Y[n,\infty]_m$ and $Z = Z[-\infty, n-1]_m$, then $[Y, Z]_{**} = 0$.*

(d) *Hence we can construct a* Postnikov tower *of slope m:*

$$\cdots \longrightarrow X[-\infty, r]_m \longrightarrow X[-\infty, r-1]_m \longrightarrow X[-\infty, r-2]_m \longrightarrow \cdots$$
$$\uparrow \qquad\qquad \uparrow \qquad\qquad \uparrow$$
$$X[r]_m \qquad\qquad X[r-1]_m \qquad\qquad X[r-2]_m$$

Here $X[r]_m$ is a generalized Eilenberg-Mac Lane spectrum: a wedge of objects of the form $\Sigma^{i,j} Hk$, where $j = mi + r$. The sequential colimit of $X[-\infty, r]_m$ is 0, and the sequential limit of $X[-\infty, r]_m$ is X.

(e) *Dually, we can construct a diagram*

$$\cdots \longrightarrow X[r,\infty]_m \longrightarrow X[r-1,\infty]_m \longrightarrow X[r-2,\infty]_m \longrightarrow \cdots$$
$$\downarrow \qquad\qquad \downarrow \qquad\qquad \downarrow$$
$$X[r]_m \qquad\qquad X[r-1]_m \qquad\qquad X[r-2]_m$$

The sequential colimit of the "connective covers" $X[r,\infty]_m$ is X, and the sequential limit is 0.

Parts (d) and (e) are easy in our context: if k is the only simple comodule, then every injective is a direct sum of copies of $\Gamma = Hk$. So to construct $X[r,\infty]_m$, for example, one just truncates the cochain complex X at bidegrees (i,j) with $j < mi + r$.

Note that if Γ is a connected Hopf algebra (Definition 0.2.2), then $\pi_{ij}S^0 = 0$ if $j < i$; since the Steenrod algebra is connected, we use this pattern for our definition of connectivity. On the other hand, if Γ is concentrated in degree zero, then $\pi_{ij}S^0 = 0$ if $j \neq 0$; this leads to a weaker notion of connectivity.

DEFINITION 1.3.4. *Given a spectrum X, if there exist numbers i_0 and j_0 so that $\pi_{ij}X = 0$ when $i < i_0$ or $j - i < j_0$, then we say that X is (i_0, j_0)-connective. We say that X is* connective *if X is (i_0, j_0)-connective for some unspecified i_0 and j_0. If for some i_0 and j_0, we have $\pi_{ij}X = 0$ when $i < i_0$ or $j < j_0$, we say that X is* weakly (i_0, j_0)-connective.

Here is the second main theorem of this section, a bigraded version of the Hurewicz theorem.

THEOREM 1.3.5. *If X is (i_0, j_0)-connective, then the Hurewicz map $\pi_{ij}X \to Hk_{ij}X$ is an isomorphism when $i < i_0$ or $j - i \leq j_0$. Similarly, if X is weakly (i_0, j_0)-connective, then $\pi_{ij}X \to Hk_{ij}X$ is an isomorphism when $i < i_0$ or $j \leq j_0$.*

If Γ is connected and X is an injective resolution of a bounded below comodule M, then X is $(0,0)$-connective, but it also satisfies a stronger property. This property occurs several times in this work, so we make it into a definition.

DEFINITION 1.3.6. We say that a spectrum X is *comodule-like*, or *CL*, if X satisfies the following conditions:
  (a) There exists an integer i_0 such that $\pi_{i*}X = 0$ if $i < i_0$,
  (b) There exists an integer j_0 such that $\pi_{ij}X = 0$ if $j - i < j_0$,
  (c) There exists an integer i_1 such that $(Hk)_{i*}X = 0$ if $i > i_1$.

For example, when X is an injective resolution of a bounded below comodule M, we may take $i_0 = i_1 = 0$, while j_0 is the degree of the bottom class of M. Note that in general, an object X is a CL-spectrum if and only if X has a cellular tower built of spheres $S^{i,j}$ with $i_0 \leq i \leq i_1$ and $j - i \geq j_0$.

We will need the following lemmas later. First we need to recall a few definitions from [**HPS97**, Definitions 1.4.3 and 2.1.1].

DEFINITION 1.3.7.   (a) A full subcategory $\mathscr{D}$ of Stable(Γ) is *localizing* if it is "closed under cofibrations and coproducts": if $X \to Y \to Z$ is a cofibration and two of X, Y, and Z are in $\mathscr{D}$, then so is the third; if $\{X_\alpha\}$ is a set of objects in $\mathscr{D}$, then $\bigvee_\alpha X_\alpha$ is in $\mathscr{D}$. Given an object Y, we let loc(Y) denote the localizing subcategory generated by Y, i.e., the intersection of all of the localizing subcategories containing Y.
  (b) Similarly, a full subcategory $\mathscr{D}$ of Stable(Γ) is *thick* if it is closed under cofibrations and retracts (if Y is in $\mathscr{D}$ and there are maps $X \to Y \to X$ so that the composite is an isomorphism, then X is in $\mathscr{D}$); and thick(Y) denotes the thick subcategory generated by Y.
  (c) A property P of spectra is *generic* if the full subcategory of spectra satisfying P is thick.
  (d) An object X of Stable(Γ) is *finite* if and only if it is small (in the categorical sense), if and only if it is in thick(S^0).

If Γ is connected, then an object X is finite if and only if X is connective and has $\dim_k Hk_{**}X < \infty$. So if Γ is connected and B is a quotient Hopf algebra of Γ, then a connective object X of Stable(Γ) is finite if and only if its restriction $X\!\downarrow_B$ is finite in Stable(B).

LEMMA 1.3.8. *Suppose that X is a spectrum and that there is a line of nonpositive slope above which the homotopy of X is zero—i.e., for some $m \leq 0$, there is an n so that if $j \geq mi + n$, then $\pi_{ij}X = 0$. Then X is in the localizing subcategory generated by $\Gamma = Hk$.*

PROOF. The tower in Theorem 1.3.3(e) displays X as being a colimit of objects of loc(Γ); hence X is itself an object of loc(Γ). $\square$

LEMMA 1.3.9. *If $\mathscr{D}$ is a localizing subcategory of Stable(Γ) which contains a nonzero finite spectrum, then* loc(Γ) $\subseteq \mathscr{D}$.

PROOF. It suffices to show that $\Gamma \in$ ob $\mathscr{D}$ if $\mathscr{D}$ is as given. Let Y be a nonzero finite object of $\mathscr{D}$; then $Y \wedge \Gamma$ is nonzero (by the Hurewicz theorem 1.3.5—remember that $\Gamma = Hk$) and is contained in ob $\mathscr{D}$. On the other hand, Corollary 1.2.10 tells us that $Y \wedge \Gamma$ is a direct sum of suspensions of Γ, so by the Eilenberg swindle [**HPS97**, Lemma 1.4.9], $\Gamma \in$ ob $\mathscr{D}$. $\square$

We will see in Corollary 6.6.8 that when $\Gamma = A$ is dual of the mod p Steenrod algebra, the containment $\mathrm{loc}(A) \subset \mathscr{D}$ is strict (i.e., $\mathrm{loc}(A)$ contains no nonzero finite spectrum).

1.4. The Adams spectral sequence

As in the rest of this chapter, we assume that Γ is a graded commutative Hopf algebra over a field k. We discuss generalized Adams spectral sequences here; we focus to some extent on the spectral sequence associated to the homology theory HB_{**}, when B is a conormal quotient Hopf algebra of Γ (see Definitions 1.2.1 and 1.2.12). In this case, we note that the Adams spectral sequence is the same as the spectral sequence associated to a Hopf algebra extension, and we mention a few consequences of this observation.

We start by giving a construction of the Adams spectral sequence, as can be found in [**Mil81**], [**Ada74**, Section III.15] or [**Rav86**, Section 2.2]. Let E be a ring spectrum satisfying the following (cf. [**Ada74**, pp. 317–318] and [**Rav86**, Assumptions 2.2.5]).

CONDITION 1.4.1. (a) E is a commutative associative ring spectrum. Let $\mu\colon E \wedge E \to E$ be the multiplication map and $\eta\colon S^0 \to E$ the unit map.
(b) E is a *flat* ring spectrum; i.e., $E_{**}E$ is flat as a left module over E_{**}.
(c) E is weakly $(0,0)$-connective (Definition 1.3.4), and the unit map η induces an isomorphism on $\pi_{0,0}$.
(d) The map $\mu_*\colon \pi_{00}E \otimes \pi_{00}E \to \pi_{00}E$ induced by the multiplication map μ is an isomorphism.
(e) $Hk_{ij}E$ is a finite-dimensional k-vector space for each i and j.

We let $\overline{E}$ denote the fiber of the unit map $S^0 \to E$. For each integer $s \geq 0$, we let
$$F_s X = \overline{E}^{\wedge s} \wedge X,$$
$$K_s X = E \wedge \overline{E}^{\wedge s} \wedge X.$$
The cofibration $\overline{E} \to S^0 \to E$ leads to the following diagram of cofibrations, which we call the (canonical) *E-based Adams tower for X*:

$$\begin{array}{ccccccccc}
X & =\!\!=\!\!= & F_0 X & \longleftarrow & F_1 X & \longleftarrow & F_2 X & \longleftarrow & \cdots \\
& & \downarrow & & \downarrow & & \downarrow & & \\
& & K_0 X & & K_1 X & & K_2 X & &
\end{array}$$

This construction satisfies the definition of an "E_*-Adams resolution" for X, as given in [**Rav86**, Definition 2.2.1]—see [**Rav86**, Lemma 2.2.9]. Note also that $F_s X = X \wedge F_s S^0$, and the same holds for $K_s X$—the canonical Adams tower is functorial and exact.

If we apply π_{**} to the Adams tower for X, we get an exact couple and hence a spectral sequence. This is called the *E-based Adams spectral sequence*.

Since we are assuming that E satisfies Condition 1.4.1, the pair $(E_{**}, E_{**}E)$ is a Hopf algebroid and $E_{**}X$ is an $E_{**}E$-comodule for any spectrum X—see [**Rav86**, Proposition 2.2.8]. As usual, we write $\mathrm{Ext}_{E_{**}E}$ for comodule Ext; since $E_{**}E$ is a bigraded coalgebra, then this Ext is trigraded.

THEOREM 1.4.2. *Let E be a spectrum satisfying Condition 1.4.1. Then there is a spectral sequence, the E-based Adams spectral sequence, which has E_2-term*
$$E_2^{s,t,u} = \operatorname{Ext}_{E_{**}E}^{s,t,u}(E_{**}, E_{**}X),$$
with differentials
$$d_r \colon E_r^{s,t,u} \to E_r^{s+r,t+r-1,u-r+1}.$$
The spectral sequence abuts to $\pi_{s+u,t+u}X$.

This theorem does not address convergence. Convergence is related to the notion of "E-completeness": a spectrum X is *E-complete* if the inverse limit of its Adams tower is contractible. If X is connective and E-complete, then the E-based Adams spectral sequence abutting to $\pi_{**}X$ will actually converge.

PROPOSITION 1.4.3. *Let E be a ring spectrum satisfying Condition 1.4.1. Then every weakly connective spectrum X is E-complete.*

PROOF. This is an easy connectivity result: our conditions on E ensure that $\overline{E}$ is weakly $(1,0)$-connective; hence if X is weakly (i,j)-connective, then by the Hurewicz theorem 1.3.5, F_sX is $(i+s,j)$-connective. □

Next we discuss the genericity of vanishing planes in Adams spectral sequences. Here is the relevant result; we do not use part (ii) of the following theorem, but we include the statement anyway.

We defined "generic" in Definition 1.3.7. If a spectrum W is (w_1, w_2)-connective (Definition 1.3.4) but neither (w_1+1, w_2)-connective nor (w_1, w_2+1)-connective, we write $\|W\| = (w_1, w_2)$. We also write $\|W\|_1$ for w_1 and $\|W\|_2$ for w_2; in other words, $\|W\|_1$ is the smallest number i so that $\pi_{i,*}W \neq 0$, and $\|W\|_2$ is the smallest number ℓ so that $\pi_{*,*+\ell}W \neq 0$.

THEOREM 1.4.4 ([**HPS99**]). *Suppose that E is a spectrum satisfying Condition 1.4.1, and consider the E-based Adams spectral sequence*
$$E_*^{***}(X) \Rightarrow \pi_{**}(X).$$
Fix numbers M and N. The following properties of an E-complete spectrum X are each generic.

(i) *There exist numbers r and b so that if s, t, and u satisfy*
$$s \geq M(s+u) + N(t+u) + b,$$
then $E_r^{s,t,u}(X) = 0$.

(ii) *There exist numbers r and b so that for all finite spectra W, if s, t, and u satisfy*
$$s \geq M(s+u-\|W\|_1) + N(t+u+\|W\|_2) + b,$$
then $E_r^{s,t,u}(X \wedge W) = 0$.

Note that the "slope" (M, N) of the vanishing plane is fixed, but the intercept b and term r of the spectral sequence may vary in these generic conditions.

REMARK 1.4.5. (a) We want E to be a nice ring spectrum so we can identify the E_2-term and so we have some convergence information. For the proof of this vanishing plane theorem, convergence is important, but the form of the E_2-term is not. Hence, we actually do not need the flatness hypothesis, Condition 1.4.1(b), and if we can guarantee convergence by some other means, then we can discard the other assumptions on E.

1.4. THE ADAMS SPECTRAL SEQUENCE

(b) Similarly, we assume that X is E-complete to ensure that the spectral sequence converges.

The proof in [**HPS99**] of the corresponding result in the ordinary stable homotopy category is formal and carries over easily to this setting.

For the remainder of this section, we let B be a conormal quotient Hopf algebra of Γ and focus on the HB-based Adams spectral sequence.

PROPOSITION 1.4.6. *Suppose that B is a conormal quotient Hopf algebra of Γ so that HB is weakly $(0,0)$-connective and so that the graded vector space $\Gamma \square_B k$ is finite-dimensional in each degree. Then HB satisfies Condition 1.4.1. Furthermore, $HB_{**}HB$ is isomorphic to $(\Gamma \square_B k) \otimes HB_{**}$ as HB_{**}-modules, and $HB_{**}HB$ is a Hopf algebra over HB_{**}.*

Note that if Γ is zero in negative dimensions, then HB will be weakly $(0,0)$-connective for any quotient B of Γ. If Γ is of finite type, then $\Gamma \square_B k$ will be, also.

PROOF. Proposition 1.2.8 says that HB is a commutative, associative ring spectrum. Flatness follows from the isomorphism $HB_{**}HB \cong (\Gamma \square_B k) \otimes HB_{**}$, which we verify below. Almost all of the rest of Condition 1.4.1 follows from the fact that, as k-algebras, $HB_{**} \cong \operatorname{Ext}_B^{**}(k,k)$; the only remaining piece is that $Hk_{ij}HB$ is finite-dimensional for each i and j. $Hk_{**}HB$ is the homology of the cochain complex HB, and so is equal to $\Gamma \square_B k$; hence our finite-dimensionality assumption on $\Gamma \square_B k$ finishes the verification of Condition 1.4.1.

To show that $HB_{**}HB$ is a Hopf algebra over HB_{**}, we first note that since B is conormal, then $\Gamma \square_B k$ is trivial as a left B-comodule. Then by Proposition 1.2.8, we have

$$HB_{**}HB = \operatorname{Ext}_B^{**}(k, \Gamma \square_B k) \cong (\Gamma \square_B k) \otimes \operatorname{Ext}_B^{**}(k,k) = (\Gamma \square_B k) \otimes HB_{**}.$$

The left unit $\eta_L \colon HB_{**} \to (\Gamma \square_B k) \otimes HB_{**}$ is the map given by applying $HB_{**}(-)$ to $k \to HB$. One can easily check that this is just the inclusion $x \mapsto 1 \otimes x$. Using the formula for the adjointness isomorphism between restriction and cotensor product in the proof of Lemma 1.2.3, one sees that η_R is the same map. Hence the Hopf algebroid $(HB_{**}, HB_{**}HB)$ is a Hopf algebra. □

For the next two results, we need to know how the gradings behave in the above isomorphism:

$$\operatorname{Ext}_B^{\ell,m}(k, \Gamma \square_B k) \cong \bigoplus_{i+j=m} (\Gamma \square_B k)_i \otimes \operatorname{Ext}_B^{\ell,j}(k,k).$$

Since $HB_{**}X$ is an $HB_{**}HB$-comodule for any X, then the proposition gives a coaction of $\Gamma \square_B k$ on $HB_{**}X$. This is the coaction mentioned in Remark 1.2.13.

COROLLARY 1.4.7. *Suppose that B is a conormal quotient of Γ as in Proposition 1.4.6. For any object $X \in \mathsf{Stable}(\Gamma)$, the E_2-term of the HB-based Adams spectral sequence abutting to $\pi_{**}X$ is isomorphic to $\operatorname{Ext}_{\Gamma \square_B k}^{**}(k, HB_{**}X)$. More precisely,*

$$E_2^{s,t,u}(X) \cong \operatorname{Ext}_{\Gamma \square_B k}^{s,t+u}(k, HB_{u,*}X).$$

PROOF. This is an exercise in homological algebra, using the above computation of $HB_{**}HB$. We leave the details to the reader. □

In the case when X is an injective resolution of a Γ-comodule M, this E_2-term is isomorphic to
$$\mathrm{Ext}^{**}_{\Gamma \square_B k}(k, \mathrm{Ext}^{**}_B(k, M)),$$
which is the E_2-term of the *change-of-rings spectral sequence* associated to the extension
$$\Gamma \square_B k \to \Gamma \to B.$$
More precisely, the change-of-rings spectral sequence has

(1.4.8) $\qquad \qquad 'E_2^{p,q,v} = \mathrm{Ext}^{p,v}_{\Gamma \square_B k}(k, \mathrm{Ext}^{q,*}_B(k, M))$

and converges (strongly) to
$$\mathrm{Ext}^{p+q,v}_\Gamma(k, M).$$
The differentials are indexed as follows:
$$d_r \colon {'E}_r^{p,q,v} \to {'E}_r^{p+r,q-r+1,v}.$$
See [**Sin73**, II, §5], for more information; alternatively, one can dualize the construction of the Lyndon-Hochschild-Serre spectral sequence for the computation of group cohomology. See [**Ben91b**, Section 3.5] for a construction of this as the spectral sequence associated to a double complex, for instance.

PROPOSITION 1.4.9. *Suppose that B is a conormal quotient of Γ, and suppose that X is an injective resolution of a Γ-comodule M. Then the HB-based Adams spectral sequence abutting to $\pi_{**}X$ is isomorphic, up to a regrading, to the change-of-rings spectral sequence associated to the extension*
$$\Gamma \square_B k \to \Gamma \to B,$$
abutting to $\mathrm{Ext}^{**}_\Gamma(k, M)$. The regrading is as follows: for all $r \geq 2$, the $E_r^{s,t,u}$-term of the Adams spectral sequence is isomorphic to the ${'E}_r^{s,u,t+u}$-term of the change-of-rings spectral sequence.

For example, the regrading gives this isomorphism of E_2-terms:
$$\mathrm{Ext}^{p,v}_{\Gamma \square_B k}(k, \mathrm{Ext}^{q,*}_B(k, M)) \cong \mathrm{Ext}^{p,v-q,q}_{HB_{**}HB}(HB_{**}, HB_{**}X).$$

PROOF. This is another exercise in homological algebra. If X is an injective resolution of a Γ-comodule M, then the Adams tower for X is a double complex; using the computation of $HB_{**}HB$ in Proposition 1.4.6, one can easily show that the resulting spectral sequence is the change-of-rings spectral sequence. □

REMARK 1.4.10. Because of this proposition, we will occasionally use the terms "HB-based Adams spectral sequence" and "change-of-rings spectral sequence" interchangeably. Note the following, however.

(a) The Adams spectral sequence abutting to $\pi_{**}X$ is defined for any object X in $\mathsf{Stable}(\Gamma)$, not just when X is an injective resolution of a comodule.
(b) On the other hand, the change-of-rings spectral sequence may be generalized somewhat: for any Γ-comodules M_1, M_2, and M_3, there is a spectral sequence with
$${'E}_2^{p,q,v} = \mathrm{Ext}^{p,v}_{\Gamma \square_B k}(M_1, \mathrm{Ext}^{q,*}_B(M_2, M_3)) \Rightarrow \mathrm{Ext}^{p+q,v}_\Gamma(M_1 \otimes M_2, M_3).$$

See [**Ben91b**, Section 3.5] for a construction of such a spectral sequence in the setting of modules over a group algebra, including a proof of convergence.

As a particular example of this spectral sequence, we have the "hypercohomology" spectral sequence: when $B = k$, then $HB = Hk = \Gamma$, so $Hk_{**}Hk = \operatorname{Ext}_k^{**}(k, \Gamma) = \Gamma$. Hence the spectral sequence has
$$E_2 \cong \operatorname{Ext}_\Gamma^{**}(k, Hk_{**}X) \Rightarrow \pi_{**}X.$$
In other words, it starts with Ext of the homology of the cochain complex X and computes the hypercohomology of X.

Because the HB-based Adams spectral sequence agrees with the change-of-rings spectral sequence, if $X = S^0$ (or more generally if X is an injective resolution of a commutative Γ-comodule algebra), then one has Steenrod operations acting on the HB-based Adams spectral sequence, as described in [**Sin73**] (see also [**Saw82**]). We need the following result in Chapter 5.

PROPOSITION 1.4.11. *Suppose that Γ is a Hopf algebra over the field $\mathbf{F}_p$, and suppose that B is a conormal quotient of Γ so that $\Gamma \square_B k$ is finite-dimensional in each degree. Consider the HB-based Adams spectral sequence converging to $\pi_{**}S^0$. Given $y \in E_2^{0,t,u}$, with t and u even if p is odd, then for each n, y^{p^n} survives to $E_{p^n+1}^{0,p^n t, p^n u}$.*

Note that the result is the same whether one is using the Adams grading—Theorem 1.4.2—or the change-of-rings grading—equation (1.4.8): the elements in $'E_2^{0,q,v}$ in the change-of-rings grading correspond to elements in $E_2^{0,v-q,q} = E_2^{0,t,u}$ in the Adams grading; those with q and v even correspond to those with t and u even.

PROOF. This follows from properties of Steenrod operations on this spectral sequence, as discussed in [**Sin73**] and [**Saw82**]. Suppose we have a spectral sequence $E_r^{s,t}$ which is a spectral sequence of algebras over the Steenrod algebra. Fix $z \in E_r^{s,t}$, and fix an integer k. If $p = 2$, then [**Sin73**, Proposition 1.4] tells us to which term of the spectral sequence $\operatorname{Sq}^k z$ survives (the result depends on r, s, t, and k). In particular, if $z \in E_r^{0,t}$, then $\operatorname{Sq}^t(z) = z^2$ survives to $E_{2r-1}^{0,2t}$. [**Saw82**, Proposition 2.5] is the corresponding result at odd primes. □

By the way, Singer's results [**Sin73**] are stated in the case of an extension of commutative Hopf algebras
$$B \to \Gamma \to C$$
where C is also cocommutative (actually, he works in the dual situation). This cocommutativity condition is not, in fact, necessary, as forthcoming work of Singer shows [**Sin99**].

1.5. Bousfield classes and Brown-Comenetz duality

Again, we suppose that Γ is a graded commutative Hopf algebra over a field k. We assume that $\mathsf{Stable}(\Gamma)$ is monogenic. In this section we collect some useful results about Bousfield classes, Brown-Comenetz duality, and their interaction.

We remind the reader that the *Bousfield class* $\langle X \rangle$ of an object X in an arbitrary stable homotopy category is the collection of X-*acyclic* objects—all objects Z with $X \wedge Z = 0$. We order Bousfield classes by reverse inclusion, so $\langle X \rangle \geq \langle Y \rangle$ means that $X \wedge Z = 0 \Rightarrow Y \wedge Z = 0$. X and Y are *Bousfield equivalent* if $\langle X \rangle = \langle Y \rangle$. We define operations $\vee$ and $\wedge$ on Bousfield classes by $\langle X \rangle \vee \langle Y \rangle = \langle X \vee Y \rangle$ and $\langle X \rangle \wedge \langle Y \rangle = \langle X \wedge Y \rangle$; one can check that these are well-defined and that $\vee$ is the least-upper bound operation. A theorem of Ohkawa [**Ohk89**] (see also [**DP**]) says that there is a set of Bousfield classes; hence there is also a greatest-lower bound operation (take the least-upper bound of the set of classes less than or equal to both $\langle X \rangle$ and $\langle Y \rangle$). For certain well-behaved classes of spectra, the greatest-lower bound is given by $\wedge$, but that is not true in general.

These concepts were introduced (in the ordinary stable homotopy category) by Bousfield in [**Bou79a**] and [**Bou79b**]; Ravenel proved a number of fundamental results about them in [**Rav84**]. See [**HPS97**, Section 3.6] for a discussion of Bousfield classes in a general stable homotopy category.

For any spectrum X, we define its *Brown-Comenetz dual* IX to be the spectrum that represents the cohomology functor

$$Y \longmapsto \mathrm{Hom}_k^{**}(\pi_{**}(X \wedge Y), k).$$

Hence $\pi_{**}IX = \mathrm{Hom}_k^{**}(\pi_{**}X, k)$. This is an analogue of the construction of Brown and Comenetz [**BC76**] in the ordinary stable homotopy category.

We say that a bigraded vector space $V = \bigoplus V_{i,j}$ is of *finite type* if $V_{i,j}$ is finite-dimensional for each i and j.

PROPOSITION 1.5.1. [**Rav84**] *Let X and Y be spectra.*
  (a) *Given a map $f: X \to X$ with cofiber X/f and telescope $f^{-1}X$, we have $\langle X/f \rangle \vee \langle f^{-1}X \rangle = \langle X \rangle$ and $\langle X/f \rangle \wedge \langle f^{-1}X \rangle = 0$.*
  (b) *If X is weakly connective (Definition 1.3.4), then IX is in the localizing subcategory generated by $Hk = \Gamma$.*
  (c) *If X has finite type homotopy, then $I^2 X = X$; hence if $[Y, X]_{**} = 0$, then $\pi_{**}(Y \wedge IX) = 0$.*
  (d) *If X is a ring spectrum and Y is an X-module spectrum, then we have $\langle Y \rangle = \langle X \wedge Y \rangle \le \langle X \rangle$.*
  (e) *If X is a ring spectrum, then IX is an X-module spectrum; hence, $\langle IX \rangle \le \langle X \rangle$.*
  (f) *Let X be a noncontractible ring spectrum with finite type homotopy and suppose that Y is an X-module spectrum. If $[Y, X]_{**} = 0$, then $\langle Y \rangle < \langle X \rangle$.*

See [**Rav84**] for the proofs. (See also Lemma 1.3.8 for (b).)

Combining part (a) of this with Lemma 1.2.15 gives us a corollary.

COROLLARY 1.5.2.   (a) *Suppose that there is a Hopf algebra extension of the form*

$$E[x] \to B \to C,$$

giving a cofiber sequence

$$\Sigma^{1,|x|}HB \xrightarrow{h} HB \to HC \to \Sigma^{0,|x|}HB.$$

Then $\langle HB \rangle = \langle HC \rangle \vee \langle h^{-1}HB \rangle$ and $HC \wedge h^{-1}HB = 0$.
  (b) *Suppose that there is a Hopf algebra extension of the form*

$$D[x] \to B \to C,$$

with $|x| > 0$, giving cofiber sequences

$$\Sigma^{2,p|x|}HB \xrightarrow{b} HB \to \widetilde{HC} \to \Sigma^{1,p|x|}HB,$$

$$\Sigma^{1,0}HC \xrightarrow{f} \Sigma^{1,|x|}HC \to \widetilde{HC} \to HC.$$

If the ring spectrum HC is weakly connective, then $\langle HB \rangle = \langle HC \rangle \vee \langle b^{-1}HB \rangle$ and $HC \wedge b^{-1}HB = 0$.

PROOF. Part (a) is immediate; part (b) follows once we show that $\langle HC \rangle = \langle \widetilde{HC} \rangle$. By the second cofibration above, it suffices to show that $f^{-1}HC$ is contractible. For connectivity reasons, though, $\pi_{**}f^{-1}HC = 0$. □

Note that $\pi_{i*}HC$ is zero when $i < 0$, so requiring HC to be weakly connective merely requires that there be a number j_0 so that $\pi_{ij}HC = \operatorname{Ext}_C^{ij}(k,k) = 0$ when $j < j_0$. This is immediate if the Hopf algebra C is zero in negative dimensions, for instance.

1.6. Further discussion

In this section, we note that we can apply the technology of this chapter to group algebras, and we give one or two examples.

Let k be a field of characteristic $p > 0$, and let G be a finite group. Then kG is a finite-dimensional cocommutative Hopf algebra, so we can use stable homotopy theory to study $\mathsf{Stable}(kG^*)$. By Remark 1.1.1, we could just as well work with the category of cochain complexes of injective kG-modules; in either case, morphisms in the category relate to group cohomology, as mentioned in Remark 1.1.2. Since kG is concentrated in degree 0, then morphisms are only singly graded.

As a result, most of the results in this chapter apply to $\mathsf{Stable}(kG^*)$. If G is p-group, then the trivial module k is the only simple, so all of the results of the chapter apply to $\mathsf{Stable}(kG^*)$ for G a p-group. To illustrate, Lemma 1.2.15 translates to the following, when $p = 2$. When B is a subgroup of G, then kB^* is a quotient Hopf algebra of kG^*; note that the coefficient ring of $H(kB^*)$ is the same as the group cohomology $H^*(B; k)$ of B.

COROLLARY 1.6.1. *Let k be a field of characteristic 2, and let G be a finite group. Any group extension*

$$1 \to B \to G \to \mathbf{Z}/2 \to 1$$

gives rise to a cofibration in $\mathsf{Stable}(kG^*)$:

$$\Sigma^1 H(kG^*) \xrightarrow{z} H(kG^*) \to H(kB^*),$$

where $z \in H^1(G; k)$ is the inflation of the polynomial generator in $H^1(\mathbf{Z}/2; k)$.

As in Lemma 1.2.15, we are using the same letter z to denote a class in $\pi_* H(kG^*)$ and the corresponding self map of $H(kG^*)$.

As mentioned in Subsection 1.2.1, when one has a group extension like

$$1 \to B \to G \to \mathbf{Z}/2 \to 1,$$

one can often replace arguments involving the associated spectral sequence with simpler arguments involving the above cofibration. It is an amusing exercise to prove Chouinard's theorem [**Cho76**] this way, for instance.

We should also discuss the Bousfield lattice in $\mathsf{Stable}(kG^*)$. In any stable homotopy category, the partially ordered set of Bousfield classes, together with the meet and join operations, contains important structural information about the category. It turns out that when G is a finite p-group, there is a complete description of the Bousfield lattice, essentially due to Benson, Carlson, and Rickard [**BCR96, BCR97**].

Given a p-group G and an algebraically closed field k of characteristic p, they define $V_G(k)$ to be the maximal ideal spectrum of the (graded) commutative Noetherian ring $H^*(G; k)$. Then for each closed homogeneous irreducible subvariety V

of $V_G(k)$, they define a kG-module $\kappa(V)$ satisfying various properties, as outlined below. Throughout, they work in the "stable module category" $\mathsf{StMod}(kG)$—see Section 3.2 for a definition of the stable comodule category, which is equivalent in this setting. This means that in their setting, $\kappa(V)$ is only well-defined up to projective summands. One could just as easily work in $\mathsf{Stable}(kG^*)$, in which case $\kappa(V)$ is a cochain complex, well-defined up to cochain-homotopy equivalence. Naturally, we choose the latter course.

THEOREM 1.6.2 ([**BCR96**]). *The objects $\kappa(V)$ satisfy the following:*
(a) *There is a Bousfield class decomposition of $\langle S^0 \rangle$:*
$$\langle S^0 \rangle = \bigvee_{V \subseteq V_G(k)} \langle \kappa(V) \rangle,$$
where the wedge is taken over all closed homogeneous irreducible V.
(b) *If $V \neq W$, then $\kappa(V) \wedge \kappa(W) = 0$.*
(c) *For any $V \subseteq V_G(k)$ and objects X and Y, if $\kappa(V) \wedge X \wedge Y = 0$, then either $\kappa(V) \wedge X = 0$ or $\kappa(V) \wedge Y = 0$.*

By [**HPS97**, Corollary 5.2.3], this implies the following.

COROLLARY 1.6.3 ([**BCR97**]). *When G is a p-group, the thick subcategories of finitely generated kG-modules are in one-to-one correspondence with collections of closed homogeneous subvarieties of $V_G(k)$ which are closed under specialization.*

Their result also has the following corollary.

COROLLARY 1.6.4. *Each $\langle \kappa(V) \rangle$ is a minimal nonzero Bousfield class. Hence the Bousfield lattice is a Boolean algebra on the classes $\langle \kappa(V) \rangle$. Indeed, for every X,*
$$\langle X \rangle = \bigvee_{V\,:\,\kappa(V) \wedge X \neq 0} \langle \kappa(V) \rangle.$$

PROOF. To show that $\langle \kappa(V) \rangle$ is minimal, we let $E = \bigvee_{W \neq V} \kappa(W)$. Then $\langle S^0 \rangle = \langle \kappa(V) \rangle \vee \langle E \rangle$, and $\kappa(V) \wedge E = 0$—in other words, $\kappa(V)$ is *complemented*. If X is any object with $\langle X \rangle < \langle \kappa(V) \rangle$, then there is a spectrum Z with $X \wedge Z = 0$ but $\kappa(V) \wedge Z \neq 0$. Hence $\kappa(V) \wedge X \wedge Z = 0$; by Theorem 1.6.2(c), this implies that $\kappa(V) \wedge X = 0$. Since $\langle S^0 \rangle = \langle \kappa(V) \rangle \vee \langle E \rangle$, then we have $\langle X \rangle = \langle X \wedge E \rangle$. On the other hand, we have $\kappa(V) \wedge E = 0$ and $\langle \kappa(V) \rangle > \langle X \rangle$, so $X \wedge E = 0$. Hence $\langle X \rangle = \langle X \wedge E \rangle = \langle 0 \rangle$, so $X = 0$.

Hence for any spectrum Y, if $\kappa(V) \wedge Y \neq 0$, then $\langle \kappa(V) \wedge Y \rangle = \langle \kappa(V) \rangle$. Combined with Theorem 1.6.2(a), this gives the desired description of the Bousfield lattice. □

These results describe much of the global structure of the category $\mathsf{Stable}(kG^*)$ (and also of $\mathsf{StMod}(kG)$), but they leave open the question of classifying all localizing subcategories (Definition 1.3.7). We conjecture that every localizing subcategory is a Bousfield class, so Corollary 1.6.4 would classify them as well. This sort of general structure is discussed in [**HPS97**, Chapter 6].

One can construct the objects $\kappa(V)$ in $\mathsf{Stable}(\Gamma)$ for any Γ whose cohomology ring is Noetherian—see [**HPS97**, Definition 6.0.8, Example 6.1.4]. This includes all finite-dimensional commutative Hopf algebras, by work of Friedlander and Suslin [**FS97**]. The analogues of parts (a) and (b) of Theorem 1.6.2 hold, but it is not clear whether part (c) does. It is natural to conjecture that a similar description of the Bousfield lattice is valid, as long as the Hopf algebra Γ is suitably well-behaved.

For instance, suppose that Γ is finite-dimensional and connected, and also that every quasi-elementary quotient of Γ is elementary (these terms are defined in Section 2.1). Hovey and the author (see [**HPb**] and [**HPa**]) were then able to imitate the work in [**BCR96**] and hence to prove Theorem 1.6.2 in this setting, giving a classification of thick subcategories and of Bousfield classes in $\mathsf{Stable}(\Gamma)$.

If the quasi-elementary quotients of Γ do not coincide with the elementaries, one still might be able to imitate Benson et al., if one had a good enough understanding of the quasi-elementary quotients of Γ. If Γ is a finite-dimensional quotient of the dual of the odd primary Steenrod algebra, the quasi-elementary quotients are not completely understood, and we do not know if Theorem 1.6.2 holds.

CHAPTER 2

Basic properties of the Steenrod algebra

The results and constructions in Chapter 1 hold as long as Γ is a graded commutative Hopf algebra over a field, occasionally with the assumption that Stable(Γ) is monogenic or that Γ is connected. Now we start to focus on the case when Γ is the dual of the Steenrod algebra.

Fix a prime p. In Section 2.1, we define the dual A of the mod p Steenrod algebra, and we give a classification of the quotient Hopf algebras of A. We also define several important families of quotient Hopf algebras of A: the $A(n)$'s, the elementary quotients, and the quasi-elementary quotients. We classify the latter two families, at least at the prime 2, and for each (quasi-)elementary E, we compute the homotopy groups of the ring spectrum HE. In Section 2.2 we introduce P_t^s-homology, a well-known tool for studying Ext over the Steenrod algebra. In our setting, P_t^s-homology is a homology theory, and hence is represented by an object P_t^s in Stable(A); we compute $\pi_{**}P_t^s$, and we perform a few other computations.

Starting in Section 2.3, we begin to get to the main results for this chapter. We discuss a vanishing line theorem in Section 2.3: given conditions on the P_t^s-homology groups of an object X, then $\pi_{ij}X = 0$ when $mi > j + c$ for some numbers m and c. (This is an extension to the cochain complex setting of theorems of Anderson-Davis [**AD73**] and Miller-Wilkerson [**MW81**].) In Section 2.4 we use the vanishing line theorem to construct "self-maps of finite objects" in Stable(A). For example, if M is an A-comodule and $\dim_{\mathbf{F}_p} M$ is finite, and if X is an injective resolution of M, then we construct a cochain map $\Sigma^{n_1,n_2}X \to X$ which is non-nilpotent under composition. We also establish that these self-maps have certain nice properties. (This is an extension to the cochain complex setting of a result of the author [**Pal92**].) In Section 2.5 we use these self-maps to construct finite objects whose P_t^s-homology groups vanish for certain specified P_t^s's, and we study some properties of these objects. We use them to study chromatic phenomena in Chapter 3.

In Section 2.6 we mention a few topological applications of the vanishing line and self-map results, and one or two other issues.

2.1. Quotient Hopf algebras of A

In this section we define the dual A of the mod p Steenrod algebra, we give a classification of quotient Hopf algebras of A, and we discuss two important families of these quotient Hopf algebras: namely, the $A(n)$'s and the elementary quotients. Margolis' book [**Mar83**] is a good reference for all of these topics. In a subsection, we also discuss the quasi-elementary quotients of A.

Fix a prime number p and let A be the dual of the mod p Steenrod algebra. Recall from [**Mil58**] Milnor's description of A: as an algebra, we have

$$A = \begin{cases} \mathbf{F}_2[\xi_1, \xi_2, \xi_3, \ldots], & \text{if } p = 2, \\ \mathbf{F}_p[\xi_1, \xi_2, \xi_3, \ldots] \otimes \Lambda[\tau_0, \tau_1, \tau_2, \ldots], & \text{if } p \text{ is odd,} \end{cases}$$

where the degrees of the generators are $|\xi_n| = 2^n - 1$ if $p = 2$, $|\xi_n| = 2(p^n - 1)$ and $|\tau_n| = 2p^n - 1$ if p odd. The coproduct map $\Delta \colon A \to A \otimes A$ is determined by its values on the generators, and then by extending multiplicatively:

$$\Delta \colon \xi_n \longmapsto \sum_{i=0}^{n} \xi_{n-i}^{p^i} \otimes \xi_i,$$

$$\Delta \colon \tau_n \longmapsto \sum_{i=0}^{n} \xi_{n-i}^{p^i} \otimes \tau_i + \tau_n \otimes 1.$$

(In both of these formulas, we take ξ_0 to be 1.)

One can check that A is a graded connected commutative Hopf algebra over the field $\mathbf{F}_p$ (or see [**Mil58**]); as a result, the work in the previous chapter applies to the study of $\mathsf{Stable}(A)$.

The quotient Hopf algebras of A have been classified by Anderson-Davis ($p = 2$) and Adams-Margolis (p odd). We state the classification theorem here; see [**Mar83**] or the original papers for the proof. (Recall that we defined conormal quotient Hopf algebras in Definition 1.2.12.)

THEOREM 2.1.1. [**AD73, AM74**]

(a) *Every quotient Hopf algebra B of A is of the form*

$$B = \begin{cases} A/(\xi_1^{2^{n_1}}, \xi_2^{2^{n_2}}, \ldots), & p = 2, \\ A/(\xi_1^{p^{n_1}}, \xi_2^{p^{n_2}}, \ldots; \tau_0^{e_0}, \tau_1^{e_1}, \ldots), & p \text{ odd,} \end{cases}$$

for some exponents $n_1, n_2, \ldots \in \{0, 1, 2, \ldots\} \cup \{\infty\}$ and $e_0, e_1, \ldots \in \{1, 2\}$. These exponents satisfy the following conditions:
  (i) *For all primes p: for each i and r with $0 < i < r$, either $n_r \geq n_{r-i} - i$ or $n_r \geq n_i$.*
  (ii) *For p odd: if $e_r = 1$, then for each i and j with $1 \leq i \leq r$ and $i + j = r$, either $n_i < j$ or $e_j = 1$.*
(b) *Conversely, any set of exponents $\{n_i\}$ and $\{e_i\}$ satisfying conditions (i)-(ii) above determines a quotient Hopf algebra of A.*
(c) *Let B be a quotient Hopf algebra of A as in (a). Then B is a conormal quotient Hopf algebra of A if and only if, for $p = 2$,*
  (i) $n_1 \leq n_2 \leq n_3 \leq \cdots$,
  and for p odd,
  (i) $n_1 \leq n_2 \leq n_3 \leq \cdots$,
  (ii) $e_0 \leq e_1 \leq e_2 \leq \cdots$,
  (iii) $e_k = 1 \Rightarrow n_k = 0$.

For part (a), if some $n_i = \infty$, that means that one does not divide out by any power of ξ_i. Similarly, if some $e_i = 2$, then one does not mod out by τ_i.

Part (a) says that there is a monomorphic function from the set of quotient Hopf algebras of A to the set of sequences either of the form $(n_1, n_2, \ldots)$, or of the form $(n_1, n_2, \ldots; e_0, e_1, \ldots)$; part (b) gives the image of this function. For $p = 2$,

2.1. QUOTIENT HOPF ALGEBRAS OF A

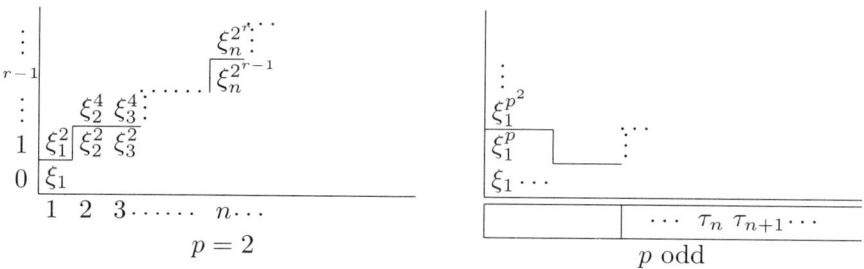

FIGURE 2.1.A. Graphical representation of a quotient Hopf algebra of A. For $p = 2$, this is a bar chart; the nth column is height $r - 1$ if one is dividing out by $\xi_n^{2^r}$. For p odd, this is a similar bar chart, together with an extra row at the bottom; in this row, one marks which τ_n's are nonzero in the quotient.

given a Hopf algebra B, one can view the sequence of exponents $n_1, n_2, \ldots$ as a function

$$\{1, 2, \ldots\} \to \{0, 1, 2, \ldots\} \cup \{\infty\},$$
$$i \longmapsto n_i.$$

We refer to this as the *profile function* of B. There is, of course, a similar function when p is odd. We will occasionally give graphical representations of quotient Hopf algebras via their profile functions, as in [**Mar83**, p. 234–5]. See Figure 2.1.A, for example.

Here is a simple, but useful, result. See Notation 0.5.1 for the definition of $D[x]$ and $E[x]$.

LEMMA 2.1.2. (a) *Suppose that B is a quotient Hopf algebra of A, and that for some s and t we have*
- $\xi_t^{p^s} \neq 0$ in B,
- $\xi_t^{p^{s+1}} = 0$ in B,
- $\xi_j^{p^s} = 0$ in B for all $j < t$.

Then there is a Hopf algebra extension

$$D[\xi_t^{p^s}] \to B \to C.$$

(b) *Fix p odd. Suppose that B is a quotient Hopf algebra of A, and that for some n we have*
- $\tau_n \neq 0$ in B, and either
- $\tau_j = 0$ in B for all $j < n$, or
- $\xi_j = 0$ in B for all $j \leq n$.

Then there is a Hopf algebra extension

$$E[\tau_n] \to B \to C.$$

PROOF. One only has to check that given the conditions on B, then $\xi_t^{p^s}$ (respectively, τ_n) is primitive in B. This check is straightforward. □

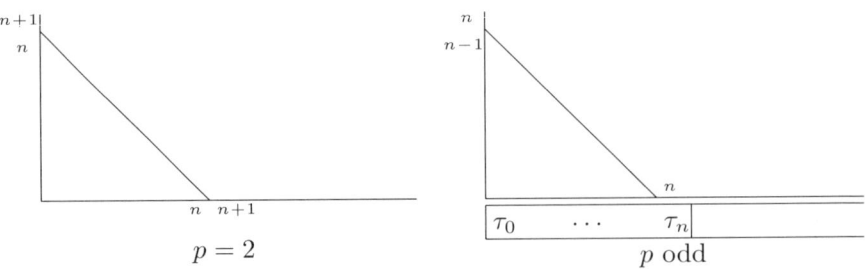

FIGURE 2.1.B. Profile functions for $A(n)$. The actual profile functions have a staircase shape, which we abbreviate as lines with slope -1.

REMARK 2.1.3. (a) Hence the results of Lemma 1.2.15 apply. The usual notation is:

$$h_{ts} = [\xi_t^{p^s}] \in HB_{1,|\xi_t^{p^s}|} = \operatorname{Ext}_B^{1,|\xi_t^{p^s}|}(\mathbf{F}_p, \mathbf{F}_p),$$
$$b_{ts} = \beta\widetilde{\mathscr{P}}^0(h_{ts}) \in HB_{2,|p\xi_t^{p^s}|} = \operatorname{Ext}_B^{2,p|\xi_t^{p^s}|}(\mathbf{F}_p, \mathbf{F}_p),$$
$$v_n = [\tau_n] \in HB_{1,|\tau_n|} = \operatorname{Ext}_B^{1,|\tau_n|}(\mathbf{F}_p, \mathbf{F}_p).$$

(b) Also, note that if B is a finite-dimensional quotient of A, then one can always find an integer n or a pair (s,t) so that the hypotheses of Lemma 2.1.2 hold. For instance, one can take s and t to be as follows:

$$t = \min\{n \mid \xi_n \neq 0 \text{ in } B\},$$
$$s = \max\{i \mid \xi_t^{p^i} \neq 0 \text{ in } B\}.$$

This provides an inductive procedure for studying finite quotients of A.

We need to use several different families of quotient Hopf algebras of A. The $A(n)$'s form the first family. These quotients are quite well-known; see [**Mar83**, p. 235], for example.

EXAMPLE 2.1.4. We define $A(n)$ as follows; see also Figure 2.1.B:

$$A(n) = \begin{cases} A/(\xi_1^{2^{n+1}}, \xi_2^{2^n}, \ldots, \xi_{n+1}^2, \xi_{n+2}, \xi_{n+3}, \ldots), & p = 2, \\ A/(\xi_1^{p^n}, \xi_2^{p^{n-1}}, \ldots, \xi_n^p, \xi_{n+1}, \xi_{n+2}, \ldots; \tau_{n+1}, \tau_{n+2}, \ldots), & p \text{ odd.} \end{cases}$$

Then $A(n)$ is a quotient Hopf algebra of A, and the map $A \to A(n)$ is an isomorphism below degree

$$|\xi_1^{2^{n+1}}| = 2^{n+1}, \quad \text{if } p = 2,$$
$$|\xi_1^{p^n}| = 2(p-1)p^n, \quad \text{if } p \text{ is odd.}$$

One important property of the $A(n)$'s is that the dual A^* of A is the union of the duals of the $A(n)$'s, and this gives A^* the structure of a "P-algebra" (see [**Mar83**, Chapter 13] for the precise definition). In our setting, this translates into the following (cf. [**Mar83**, Proposition 13.4]).

PROPOSITION 2.1.5. *Suppose that Y is a spectrum so that for each i, $\pi_{ij}Y = 0$ when $j \ll 0$. Let B be a quotient Hopf algebra of A, and define $B(n)$ by the following* *pushout diagram of Hopf algebras:*

$$\begin{array}{ccc} A & \twoheadrightarrow & B \\ \downarrow & & \downarrow \\ A(n) & \twoheadrightarrow & B(n). \end{array}$$

Then $HB \wedge Y$ is the sequential limit of

$$\cdots \to HB(3) \wedge Y \to HB(2) \wedge Y \to HB(1) \wedge Y \to HB(0) \wedge Y.$$

Hence for any spectrum X, there is a Milnor exact sequence

$$0 \to \varprojlim{}^1 [X, HB(n) \wedge Y]_{i-1,j} \to [X, HB \wedge Y]_{i,j} \to \varprojlim [X, HB(n) \wedge Y]_{i,j} \to 0.$$

Here are two alternate descriptions of $B(n)$. First, the profile function for $B(n)$ is given by intersecting the profile functions of B and $A(n)$. Second, in the setting of sub-Hopf algebras of the dual of A, the dual of $B(n)$ is $B(n)^* = B^* \cap A(n)^*$.

PROOF. We write the cochain complex Y as $\cdots \to Y_j \to Y_{j+1} \to \cdots$. We may assume that for each j, the injective comodule Y_j is bounded below. Because $A \twoheadrightarrow A(n)$ is an isomorphism in a range of dimensions increasing with n, so is the map $B \twoheadrightarrow B(n)$; hence the inverse system of comodules

$$\cdots \to (A \,\square_{B(n)} \mathbf{F}_p) \otimes Y_j \to (A \,\square_{B(n-1)} \mathbf{F}_p) \otimes Y_j \to (A \,\square_{B(n-2)} \mathbf{F}_p) \otimes Y_j \to \cdots$$

stabilizes in any given degree, and the inverse limit is $(A \,\square_B \mathbf{F}_p) \otimes Y_j$. We let $(A \,\square_{B(n)} \mathbf{F}_p) \otimes Y$ denote the cochain complex which is $(A \,\square_{B(n)} \mathbf{F}_p) \otimes Y_j$ in degree j; then the inverse system of cochain complexes stabilizes in each bidegree, and the inverse limit is $HB \wedge Y$. This finishes the proof. (Note that $(A \,\square_{B(n)} \mathbf{F}_p) \otimes Y$ is isomorphic to $HB(n) \wedge Y$ in $\mathsf{Stable}(A)$.) □

For example, with Y as in the proposition and $B = A$, Y is the sequential limit of

$$\cdots \to HA(3) \wedge Y \to HA(2) \wedge Y \to HB(1) \wedge Y \to HA(0) \wedge Y.$$

We consider one other family of quotient Hopf algebras. See [**Wil81**] for some results related to these.

DEFINITION 2.1.6. We say that a connected commutative Hopf algebra B over a field k of characteristic p is *elementary* if it is isomorphic (as a Hopf algebra) to a tensor product of Hopf algebras of the forms $k[x]/(x^{p^n})$ with x primitive (and $|x|$ even, if p is odd), and $E[y]$ with y primitive (and $|y|$ odd, if p is odd). In other words, B is bicommutative and its dual B^* has $z^p = 0$ for all z in the augmentation ideal.

See [**Mar83**], [**Lin77a**] and [**Wil81**] for the following.

PROPOSITION 2.1.7. (a) *Suppose that $p = 2$. A quotient Hopf algebra*

$$B = A/(\xi_1^{2^{n_1}}, \xi_2^{2^{n_2}}, \ldots)$$

of A is elementary if and only if for some r,
  (i) *if $i < r$, then $n_i = 0$,*
  (ii) *if $i \geq r$, then $n_i \leq r$.*

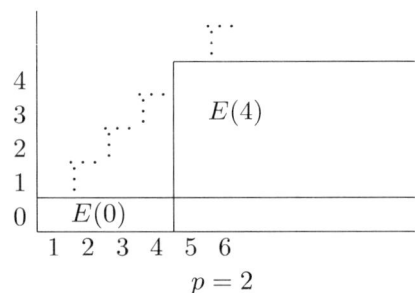

FIGURE 2.1.C. Profile functions for maximal elementary quotients of A at the prime 2. For reference, we have included a staircase above which are the elements $\xi_t^{2^s}$ with $s \geq t$.

(b) *Suppose that p is odd. A quotient Hopf algebra*
$$B = A/(\xi_1^{p^{n_1}}, \xi_2^{p^{n_2}}, \ldots; \tau_0^{e_0}, \tau_1^{e_1}, \ldots),$$
of A is elementary if and only if either $n_i = 1$ for all i, or for some r,
  (i) *if $i < r$, then $n_i = 0$,*
  (ii) *if $i \geq r$, then $n_i \leq r$,*
  (iii) *if $i < r$, then $e_i = 1$.*

We also describe the maximal elementary quotient Hopf algebras of A; see Figure 2.1.C for the profile function at the prime 2. Every elementary quotient Hopf algebra of A is a quotient of one of these.

COROLLARY 2.1.8. (a) *Suppose that $p = 2$. The maximal elementary quotient Hopf algebras of A are*
$$E(m) = A/(\xi_1, \ldots, \xi_m, \xi_{m+1}^{2^{m+1}}, \xi_{m+2}^{2^{m+1}}, \xi_{m+3}^{2^{m+1}}, \ldots), \ m \geq 0.$$

(b) *Suppose that p is odd. The maximal elementary quotient Hopf algebras of A are*
$$E(-1) = A/(\xi_1, \xi_2, \ldots) \cong \Lambda[\tau_0, \tau_1, \ldots],$$
$$E(m) = A/(\xi_1, \ldots, \xi_m, \xi_{m+1}^{p^{m+1}}, \xi_{m+2}^{p^{m+1}}, \xi_{m+3}^{p^{m+1}}, \ldots; \tau_0, \ldots, \tau_m), \ m \geq 0.$$

Note that the quotient Hopf algebras $E(m)$ are conormal for all m.

We use the elementary quotient Hopf algebras of A to prove the vanishing line theorem of Section 2.3; we need to know their coefficient rings. The following is standard; it also follows from Proposition 0.5.10 and the Künneth theorem.

PROPOSITION 2.1.9. (a) *Let $p = 2$, and assume that E is an elementary quotient Hopf algebra of A. Then*
$$HE_{**} = \mathbf{F}_2[h_{ts} \mid \xi_t^{2^s} \neq 0 \text{ in } E].$$
Here, $|h_{ts}| = (1, |\xi_t^{2^s}|)$.

(b) *Let p be odd, and assume that E is an elementary quotient Hopf algebra of A. Then*
$$HE_{**} = \Lambda[h_{ts} \mid \xi_t^{p^s} \neq 0 \text{ in } E] \otimes \mathbf{F}_p[b_{ts} \mid \xi_t^{p^s} \neq 0 \text{ in } E]$$
$$\otimes \mathbf{F}_p[v_n \mid \tau_n \neq 0 \text{ in } E].$$
Here, $|h_{ts}| = (1, |\xi_t^{p^s}|)$, $|b_{ts}| = (2, p|\xi_t^{p^s}|)$, and $|v_n| = (1, |\tau_n|)$.

Indeed, if E is an elementary quotient of A with $\xi_t^{p^s} \neq 0$, then $\xi_t^{p^s}$ is primitive in E. Following Notation 1.2.14 and Remark 2.1.3(a), we have $h_{ts} = [\xi_t^{p^s}]$; similarly, if τ_n is non-zero in E, then it is primitive and $v_n = [t_n]$.

2.1.1. Quasi-elementary quotients of A. We also need to consider one other family of quotient Hopf algebras, the "quasi-elementary" Hopf algebras. It turns out that when $p = 2$, these coincide with elementary Hopf algebras; when p is odd, there are quasi-elementary Hopf algebras which are not elementary, but we do not have a complete classification.

The quasi-elementary quotients of A arise in the nilpotence theorem 5.1.6 of Section 5.1, and indeed in many of the results of Chapters 5 and 6; the lack of a classification at odd primes is one obstacle to proving those theorems when p is odd. So when p is 2, we have already studied these; when p is odd, we have no immediate use for them. Hence the contents of this subsection may be safely ignored, except for the term "quasi-elementary."

DEFINITION 2.1.10. Let B be a connected commutative Hopf algebra B over a field k.
  (a) Suppose that k has characteristic 2. Then B is *quasi-elementary* if for every finite set $S \subset \operatorname{Ext}_B^{1,*}(k,k) - \{0\}$, the product $\prod_{w \in S} w$ is non-nilpotent.
  (b) Suppose that k has characteristic $p > 2$. Recall from Notation 1.2.14 that there is a Steenrod operation $\beta \widetilde{\mathscr{P}}^0$ on $\operatorname{Ext}_B^{**}(k,k)$. B is *quasi-elementary* if for all finite sets
$$S_{\text{odd}} \subset \bigoplus_{n \in \mathbf{Z}} \operatorname{Ext}_B^{1, 2n+1}(k,k) - \{0\},$$
$$S_{\text{even}} \subset \bigoplus_{n \in \mathbf{Z}} \operatorname{Ext}_B^{1, 2n}(k,k) - \{0\},$$
the product
$$(\prod_{u \in S_{\text{odd}}} u)(\prod_{v \in S_{\text{even}}} \beta \widetilde{\mathscr{P}}^0 v)$$
is non-nilpotent.

Every elementary Hopf algebra is quasi-elementary; Wilkerson [**Wil81**, Section 6] gives several examples of quasi-elementary Hopf algebras which are not elementary. In particular, one of his examples is a quotient of A, when p is odd. See [**Wil81**, Counterexample 6.3] for the following.

EXAMPLE 2.1.11. Suppose that p is odd, and let B be this quotient Hopf algebra of A:
$$B = \mathbf{F}_p[\xi_1, \xi_2, \xi_3]/(\xi_1^p, \xi_2^{p^2}, \xi_3^p).$$
Then B is quasi-elementary, but not elementary.

When p is 2, things are a bit nicer. See [**Wil81**, Theorem 6.4] for the following.

PROPOSITION 2.1.12. *Suppose that* $p = 2$. *A quotient Hopf algebra* B *of* A *is elementary if and only if it is quasi-elementary.*

Note that Lin [**Lin77a**] has classified the elementary quotients of A—see Proposition 2.1.7 and Corollary 2.1.8 above. We have also computed the coefficient rings of these Hopf algebras in Proposition 2.1.9.

At odd primes, we do not have a classification of the quasi-elementary Hopf algebras, so we content ourselves with the following. This would follow from Conjecture 5.4.1.

CONJECTURE 2.1.13. *Suppose that p is odd. Then every quasi-elementary quotient Hopf algebra of A is a quotient of*

$$D = A/(\xi_1^p, \xi_2^{p^2}, \xi_3^{p^3}, \ldots, \xi_n^{p^n}, \ldots).$$

See [**Pal97**] for some results relating to quasi-elementary Hopf algebras. We should also point out that there is a classification of finite-dimensional quasi-elementary quotients of A at odd primes in [**NP98**]; however, the proof given there has a gap.

2.2. P_t^s-homology

In this section we discuss P_t^s-homology; this tool has been used by many authors to study Ext_A, so it is reasonable to expect it to be useful in the present setting. [**Mar83**, Section 19.1] is a good reference for basic results on P_t^s-homology of modules; in a subsection, we prove analogues of some of those results.

Let A^* be the dual of A. Note that an A-comodule M with structure map ψ is naturally an A^*-module, via the composite

$$A^* \otimes M \xrightarrow{1 \otimes \psi} A^* \otimes A \otimes M \xrightarrow{\operatorname{ev} \otimes 1} M.$$

We remind the reader that if one dualizes with respect to the monomial basis for A, then P_t^s is the element of A^* dual to $\xi_t^{p^s}$, and (when p is odd) Q_n is the element dual to τ_n. If $s < t$, then $(P_t^s)^p = 0$, so in this case one may define the P_t^s-homology of an A-comodule M by

$$H(M, P_t^s) = \frac{\ker(M \xrightarrow{(P_t^s)^{p-1}} M)}{\operatorname{im}(M \xrightarrow{P_t^s} M)}.$$

For all n we have $Q_n^2 = 0$, so one may define $H(M, Q_n)$ similarly. One *may* do these things, but that does not mean that one *should* do them when working in the context of cochain complexes of injective A-comodules.

We defined the coalgebras $E[x]$ and $D[x]$ in Notation 0.5.1.

DEFINITION 2.2.1. Given integers s and t with $0 \leq s < t$, we define the *connective P_t^s-homology spectrum*, p_t^s, to be the cochain complex $HD[\xi_t^{p^s}]$. In other words, it is an injective resolution of $A \,\square_{D[\xi_t^{p^s}]} \mathbf{F}_p$. Hence it is chain homotopy equivalent to the complex obtained by applying $A \,\square_{D[\xi_t^{p^s}]} -$ to the injective resolution of the $D[\xi_t^{p^s}]$-comodule $\mathbf{F}_p$ as given in the diagram (0.5.9). So for $j \geq 0$, we may take its jth term to be

$$(p_t^s)_j = \begin{cases} s^{kp|\xi_t^{p^s}|}A, & j = 2k, \\ s^{(kp+1)|\xi_t^{p^s}|}A, & j = 2k+1, \end{cases}$$

where s denotes the "internal" suspension functor on graded comodules. The differentials are given as in the following diagram:

$$\cdots \longrightarrow A \longrightarrow A \longrightarrow A \longrightarrow A \longrightarrow \cdots.$$
$$\xi_t^{p^s} \longmapsto 1, \quad \xi_t^{(p-1)p^s} \longmapsto 1, \quad \xi_t^{p^s} \longmapsto 1$$

(A is a rank 1 cofree comodule, so a comodule map to A is determined by specifying a vector space map to $\mathbf{F}_p$—see Lemma 0.4.3. In this diagram, "$x \mapsto 1$" indicates the vector space map in which x goes to $1 \in \mathbf{F}_p$, and the other elements in the monomial basis for A go to 0. One can also describe this map as "right multiplication by x^*," where $x^* \in A^*$ is the element dual to $x \in A$.) Note that this cochain complex is periodic with period 2 if p is odd, period 1 if $p = 2$. We define the (periodic) P_t^s-homology spectrum, P_t^s, to be the cochain complex obtained by extending the periodicity to negative dimensions; in other words, for all integers j,

$$(P_t^s)_j = \begin{cases} s^{kp|\xi_t^{p^s}|}A, & j = 2k, \\ s^{(kp+1)|\xi_t^{p^s}|}A, & j = 2k+1, \end{cases}$$

with differentials given as above. In other words, this cochain complex is obtained by applying $A \square_{D[\xi_t^{p^s}]} -$ to the long exact sequence of $D[\xi_t^{p^s}]$-comodules

$$\cdots \to s^{kp|\xi_t^{p^s}|}D[\xi_t^{p^s}] \to s^{(kp+1)|\xi_t^{p^s}|}D[\xi_t^{p^s}] \to s^{(k+1)p|\xi_t^{p^s}|}D[\xi_t^{p^s}] \to \cdots.$$

Similarly, when p is odd, we define the *connective Q_n-homology spectrum*, q_n, to be the cochain complex $HE[\tau_n]$. Hence it may be obtained by applying $A \square_{E[\tau_n]} -$ to the long exact sequence (0.5.7), in which case its jth term is given by

$$(q_n)_j = \begin{cases} 0 & \text{if } j < 0, \\ s^{j|\tau_n|}A & \text{if } j \geq 0, \end{cases}$$

with differentials

$$\cdots \longrightarrow A \longrightarrow A \longrightarrow A \longrightarrow A \longrightarrow \cdots.$$
$$\tau_n \longmapsto 1, \; \tau_n \longmapsto 1, \; \tau_n \longmapsto 1$$

We define the (periodic) Q_n-homology spectrum, Q_n, to be the periodic version of q_n; in other words, for all integers j, $(Q_n)_j = s^{j|\tau_n|}A$, with differentials as above. For any spectrum X, we define its P_t^s-homology to be $(P_t^s)_{**}X$, and we define its Q_n-homology to be $(Q_n)_{**}X$.

From the construction, we can compute the homology $(H\mathbf{F}_p)_{**}$ of these spectra.

LEMMA 2.2.2.   (a) For all $s < t$,

$$(H\mathbf{F}_p)_{i,*}p_t^s = \begin{cases} A \square_{D[\xi_t^{p^s}]}\mathbf{F}_p, & \text{if } i = 0, \\ 0, & \text{if } i \neq 0. \end{cases}$$

(b) For all $s < t$, $(H\mathbf{F}_p)_{**}P_t^s = 0$.
(c) For all n,

$$(H\mathbf{F}_p)_{i,*}q_n = \begin{cases} A \square_{E[\tau_n]}\mathbf{F}_p, & \text{if } i = 0, \\ 0, & \text{if } i \neq 0. \end{cases}$$

(d) For all n, $(H\mathbf{F}_p)_{**}Q_n = 0$.

PROOF. p_t^s is an injective resolution of $A \square_{D[\xi_t^{p^s}]} \mathbf{F}_p$ so it only has homology in degree zero. P_t^s is defined by extending the periodicity of the p_t^s cochain complex into negative dimensions, so it has no homology at all. The same arguments work for q_n and Q_n. □

In the remainder of this section, we compute the coefficient rings of these spectra and we prove a few results for later use.

PROPOSITION 2.2.3. (a) If $s < t$, then $D[\xi_t^{p^s}]$ is a quotient coalgebra of A over which A is injective as a right comodule. For any $n \geq 0$, $E[\tau_n]$ is a quotient Hopf algebra of A, and hence A is injective as a right comodule over it.

(b) Hence,
$$(p_t^s)_{**} \cong \begin{cases} \mathbf{F}_2[h_{ts}], & p = 2, \\ \mathbf{F}_p[b_{ts}] \otimes \Lambda[h_{ts}], & p \text{ odd}, \end{cases}$$
where $|h_{ts}| = (1, |\xi_t^{p^s}|)$ and $|b_{ts}| = (2, p|\xi_t^{p^s}|)$. Also,
$$(q_n)_{**} \cong \mathbf{F}_p[v_n],$$
where $|v_n| = (1, |\tau_n|)$.

PROOF. Part (a) is well-known. It is dual to the statement that A^*, the dual of A, is free over the subalgebra $\mathbf{F}_p[P_t^s]/(P_t^s)^p$; i.e., $H(A^*, P_t^s) = 0$. (And similarly, A^* is free over $E[Q_n]$; i.e., $H(A^*, Q_n) = 0$.) See [**AM71**], [**MP72**], or [**Mar83**, Proposition 19.1].

Part (b) is standard, trivial, or both, using Propositions 0.5.10 and 1.2.8, for instance. □

By construction, the periodicity of the cochain complex p_t^s yields a self-map of one of the following forms:
$$h_{ts} \colon \Sigma^{1, |\xi_t^{2^s}|} p_t^s \to p_t^s, \quad \text{if } p = 2,$$
$$b_{ts} \colon \Sigma^{2, p|\xi_t^{p^s}|} p_t^s \to p_t^s, \quad \text{if } p \text{ is odd}.$$

On homotopy groups, the first of these induces multiplication by the element $h_{ts} \in \pi_{1, |\xi_t^{2^s}|}(p_t^s)$, and similarly for the second; this explains our names for the self-maps at the cochain complex level. At odd primes, q_n has a self-map $v_n \colon \Sigma^{1, |\tau_n|} q_n \to q_n$. Taking the mapping telescopes of these self-maps gives the periodic spectra P_t^s and Q_n, respectively; indeed, one can take this to be the definition of those spectra.

LEMMA 2.2.4. (a) Let $p = 2$. For all $s < t$, P_t^s is the mapping telescope of the self-map $h_{ts} \colon p_t^s \to \Sigma^{-1, -|\xi_t^{2^s}|} p_t^s$. In symbols, $P_t^s = h_{ts}^{-1} p_t^s$.
(b) Let p be an odd prime. For all $s < t$, $P_t^s = b_{ts}^{-1} p_t^s$.
(c) Let p be an odd prime. For all n, $Q_n = v_n^{-1} q_n$.

PROOF. This follows from the definition of P_t^s and Q_n in terms of p_t^s and q_n. □

See [**Mar83**, Propositions 19.2–3] for an alternate formulation of the following.

COROLLARY 2.2.5. Hence
$$(P_t^s)_{**} = \begin{cases} \mathbf{F}_2[h_{ts}^{\pm 1}], & p = 2, \\ \mathbf{F}_p[b_{ts}^{\pm 1}] \otimes \Lambda[h_{ts}], & p \text{ odd}, \end{cases}$$
$$(Q_n)_{**} = \mathbf{F}_p[v_n^{\pm 1}],$$
and for any connective X,
$$(P_t^s)_{**} X = \begin{cases} h_{ts}^{-1} (p_t^s)_{**} X, & p = 2, \\ b_{ts}^{-1} (p_t^s)_{**} X, & p \text{ odd}, \end{cases}$$
$$(Q_n)_{**} X = v_n^{-1} (q_n)_{**} X.$$

By the way, we note that for all primes, the spectrum $p_t^0 = HD[\xi_t]$ is a ring spectrum, while P_t^0 is a field spectrum for $p = 2$ because P_t^s is commutative and every homogeneous element in $\pi_{**}P_t^s$ is invertible—see [**HPS97**, Proposition 3.7.2]. (Perhaps P_t^0 is an example of what one might call an "Artinian ring spectrum" when p is odd.) Similarly, when p is odd, $q_n = HE[\tau_n]$ is a ring spectrum and Q_n is a field spectrum. The spectra p_t^s for $s > 0$ are not ring spectra, but are examples of the sort discussed in Proposition 1.2.8(b).

REMARK 2.2.6. As one might expect, there is a relationship between the module definition of P_t^s-homology (given just before Definition 2.2.1) and the homology functor represented by the P_t^s-homology spectrum. This relationship is straightforward when $p = 2$, more complicated when p is odd. If X is an injective resolution of a comodule M of finite-type, then the change-of-rings isomorphism of Example 1.2.4 gives
$$(p_t^s)_{**}X \cong \operatorname{Ext}^{**}_{D[\xi_t^{p^s}]}(\mathbf{F}_p, M).$$
Corollary 0.5.5 gives a classification of finite-type $D[\xi_t^{p^s}]$-comodules, so we just need to know these Ext groups for each indecomposable summand. These are easily computed by taking minimal resolutions; the answers are given in Proposition 0.5.10. When $p = 2$, the indecomposable comodules are the trivial comodule $\mathbf{F}_p$ and the cofree comodule. We have the following isomorphism for each of these, and hence for every comodule M: when $i \geq 1$,

(2.2.7) $\qquad \operatorname{Ext}^{i,*}_{D[\xi_t^{p^s}]}(\mathbf{F}_p, M) \cong H_*(M, P_t^s) \otimes \operatorname{Ext}^{i,*}_{D[\xi_t^{p^s}]}(\mathbf{F}_p, \mathbf{F}_p).$

Hence the periodic P_t^s-homology of the injective resolution X of M is given by
$$(P_t^s)_{**}X = H_*(M, P_t^s) \otimes (P_t^s)_{**}.$$
(Here we are viewing the singly-graded vector space $H_*(M, P_t^s)$ as doubly-graded by putting $H_j(M, P_t^s)$ in bidegree $(0, j)$.) In particular, $(P_t^s)_{**}X = 0$ if and only if $H_*(M, P_t^s) = 0$.

Similarly, we have
$$(Q_n)_{**}X = H_*(M, Q_n) \otimes (Q_n)_{**}.$$
In the P_t^s case when p is odd, the isomorphism (2.2.7) is not quite valid for every indecomposable comodule M; rather, one finds that
$$\operatorname{Ext}^{i,*}_{D[\xi_t^{p^s}]}(\mathbf{F}_p, M) \cong \begin{cases} s^{e(M)} H_*(M, P_t^s) \otimes \operatorname{Ext}^{i,*}_{D[\xi_t^{p^s}]}(\mathbf{F}_p, \mathbf{F}_p), & \text{if } i \text{ is even,} \\ s^{o(M)} H_*(M, P_t^s) \otimes \operatorname{Ext}^{i,*}_{D[\xi_t^{p^s}]}(\mathbf{F}_p, \mathbf{F}_p), & \text{if } i \text{ is odd,} \end{cases}$$
where s is the suspension functor in the category of graded vector spaces, and $e(M)$ and $o(M)$ are numbers that depend on the indecomposable comodule M. In particular, we still see that if M is any finite-type comodule and X is its injective resolution, $(P_t^s)_{**}X = 0$ if and only if $H_*(M, P_t^s) = 0$.

It is convenient to have alternate notation for the spectra P_t^s and Q_n, based on the "slopes" of the polynomial generators in their coefficient rings. We use this notation for much of the remainder of the book.

NOTATION 2.2.8. We define the *slope* of P_t^s to be $\frac{p|\xi_t^{p^s}|}{2}$, and the slope of Q_n to be $|\tau_n|$. Note that these spectra all have distinct slopes. The set

$$\{|\xi_t^{2^s}| : s < t\}, \qquad p = 2,$$
$$\{\frac{p|\xi_t^{p^s}|}{2} : s < t\} \cup \{|\tau_n| : n \geq 0\}, \quad p \text{ odd},$$

is called the set of *slopes* of A; the phrase "fix a slope m" means "fix an element m of this set." Given a slope m, we let $Z(m)$ denote the corresponding P_t^s or Q_n spectrum. For example, when $p = 2$, we have

$$Z(1) = P_1^0, \ Z(3) = P_2^0, \ Z(6) = P_2^1, \ Z(7) = P_3^0, \ \ldots, \ Z(|\xi_t^{2^s}|) = P_t^s, \ \ldots.$$

When p is odd, we have

$$Z(1) = Q_0, \ Z(p-1) = Q_1, \ Z(p^2-p) = P_1^0, \ Z(p^2-1) = Q_2, \ \ldots,$$

$$Z(\frac{p|\xi_t^{p^s}|}{2}) = P_t^s, \ \ldots, \ Z(|\tau_n|) = Q_n, \ \ldots.$$

Note that $Z(4)$, for instance, is meaningless when $p = 2$, because 4 is not a slope at that prime. We let $z(m)$ be the connective version of $Z(m)$, either p_t^s or q_n. We let y_m denote the corresponding element of A, either $\xi_t^{p^s}$ or τ_n.

2.2.1. Miscellaneous results about P_t^s-homology. In this subsection, we present a few results about P_t^s-homology, either for later use or for comparison with work of other authors, especially Margolis.

See [**Mar83**, Theorem 19.21] for the following at the prime 2.

PROPOSITION 2.2.9. *Let B be a quotient Hopf algebra of A. Then $(P_t^s)_{**}HB \neq 0$ if and only if $\xi_t^{p^s} \neq 0$ in B. For p odd, $(Q_n)_{**}HB \neq 0$ if and only if $\tau_n \neq 0$ in B.*

We will use this result in Section 3.3.

PROOF. We will give the proof for P_t^s and leave the Q_n proof for the reader. By Remark 2.2.6, we are interested in $H(A \square_B \mathbf{F}_p, P_t^s)$. If $\xi_t^{p^s} \neq 0$ in B, then $1 \in A \square_B \mathbf{F}_p$ generates a nonzero homology class.

To prove the converse, first we reduce to the case when B is finite. We define $B(n)$ to be the quotient Hopf algebra of A defined as in Proposition 2.1.5 by the following pushout diagram of Hopf algebras:

$$\begin{array}{ccc} A & \longrightarrow & B \\ \downarrow & & \downarrow \\ A(n) & \longrightarrow & B(n). \end{array}$$

($A(n)$ is defined in Example 2.1.4.) In other words, $B(n)$ is the quotient of B induced by the map $A \twoheadrightarrow A(n)$. Note that each $B(n)$ is finite, and if $\xi_t^{p^s} = 0$ in B, then the same is true in $B(n)$ for all n. So if we assume the result for finite quotients of A, then we know that $(P_t^s)_{**}HB(n) = 0$ for all n. Hence the Milnor short exact sequence in Proposition 2.1.5

$$0 \to \varprojlim{}^1[S^0, HB(n) \wedge P_t^s]_{i-1,j} \to [S^0, HB \wedge P_t^s]_{i,j} \to \varprojlim[S^0, HB(n) \wedge P_t^s]_{i,j} \to 0$$

has first and last terms zero; therefore the middle term is also zero.

So it suffices to show that if B is a finite quotient of A in which $\xi_t^{p^s} = 0$, then $(P_t^s)_{**}HB = 0$. We do this by induction on $\dim_{\mathbf{F}_p} B$. The induction starts with $B = \mathbf{F}_p$, in which case $H\mathbf{F}_p = A$. Then
$$(P_t^s)_{**}H\mathbf{F}_p = (H\mathbf{F}_p)_{**}P_t^s,$$
the homology of the cochain complex P_t^s. This complex is acyclic, so its homology is zero—see Lemma 2.2.2. (Alternatively, see the remark following this proof.) This starts the induction.

Now fix a finite-dimensional B with $\xi_t^{p^s} = 0$ in B, and assume that for every proper quotient C of B, we have $(P_t^s)_{**}HC = 0$. As noted in Remark 2.1.3, there is a Hopf algebra quotient C of B with Hopf algebra kernel either $\mathbf{F}_p[\xi_r^{p^q}]/(\xi_r^{p^{q+1}})$ or $E[\tau_m]$. So by Lemma 1.2.15 this leads to a cofiber sequence (in which we neglect suspensions)
$$HB \xrightarrow{f} HB \to Z,$$
where either $Z = HC$ or Z is the cofiber of a self-map of HC. In either case, $(P_t^s)_{**}HC = 0 \Rightarrow (P_t^s)_{**}Z = 0$, so $(P_t^s)_{**}f$ is an isomorphism.

Now we argue essentially as in the proof of Lemma 2.3.12 to see that because comodule $A\square_B\mathbf{F}_p$ is bounded below, then $(P_t^s)_{**}HB = 0$. To be precise, we first have to specify the degree of the map f. There are three cases:

(1) $p = 2$: then f has bidegree $(1, |\xi_r^{2^q}|)$—i.e., f is a map $\Sigma^{1,|\xi_r^{2^q}|}HB \to HB$. ($f$ is also known as h_{rq}.)
(2) p odd, Hopf algebra kernel $E[\tau_m]$: then f has bidegree $(1, |\tau_m|)$. (f is also known as v_m.)
(3) p odd, Hopf algebra kernel $D[\xi_r^{p^q}] = \mathbf{F}_p[\xi_r^{p^q}]/(\xi_r^{p^{q+1}})$: then f has bidegree $(2, p|\xi_r^{p^q}|)$. (f is also known as b_{rq}.)

We deal with cases (1) and (3); (2) is handled the same way. By our computations in Corollary 2.2.5, we know that for all i and j, we have an isomorphism
$$(P_t^s)_{ij}HB \cong (P_t^s)_{i+2, j+p|\xi_t^{p^s}|}HB.$$
(This is given by $(h_{ts})^2$ when $p = 2$, and by b_{ts} when p is odd.) In cases (1) and (3), f induces an isomorphism
$$(P_t^s)_{ij}HB \cong (P_t^s)_{i+2, j+p|\xi_r^{p^q}|}HB.$$
(In other words, f^2 when $p = 2$, and f when p is odd.) Combining these, for each integer k we get an isomorphism
$$(P_t^s)_{ij}HB \cong (P_t^s)_{i, j+pk(|\xi_t^{p^s}|-|\xi_r^{p^q}|)}HB.$$

Now, $\xi_r^{p^q} \neq 0$ in B and $\xi_t^{p^s} = 0$, so $\xi_r^{p^q} \neq \xi_t^{p^s}$. In particular, these two elements have different degrees. By Remark 2.2.6, for fixed i, $(P_t^s)_{i,*}HB$ is equal (up to suspension) to $H_*(A\square_B\mathbf{F}_p, P_t^s)$; since $A\square_B\mathbf{F}_p$ is a bounded below comodule, then this homology must be zero in small enough degrees. By the above isomorphism, we may then conclude that it is zero in all degrees. $\square$

We remark that it is easy to show that $(H\mathbf{F}_p)_{**}P_t^s = (P_t^s)_{**}A = 0$—this follows by a result of Milnor-Moore [**MM65**], as in [**Mar83**, Proposition 19.1]. It seems as though there should be a similar proof of Proposition 2.2.9, but we have not been able to find one.

Recall from [**HPS97**, Definition A.2.4] that if X is any object in $\mathsf{Stable}(A)$, then DX denotes the *Spanier-Whitehead dual* of X. DX is defined to be the object representing the cohomology theory $Y \longmapsto [X \wedge Y, S^0]$; hence for any Y,
$$[X \wedge Y, S^0] = [Y, DX].$$
We mention the following easy fact.

PROPOSITION 2.2.10. *Let E and X be spectra with X finite. Then $E_{**}X = 0$ if and only if $E_{**}DX = 0$.*

PROOF. If X is E-acyclic, then so is $X \wedge Y$ for any Y. In particular, $DX \wedge X \wedge DX$ is E-acyclic. Since DX is a retract of this when X is finite, we are done. □

COROLLARY 2.2.11. *Let X be a finite spectrum. Then $(P_t^s)_{**}X = 0$ if and only if $(P_t^s)_{**}DX = 0$; and for p odd, $(Q_n)_{**}X = 0$ if and only if $(Q_n)_{**}DX = 0$.*

One should be able to get more precise information about the relationship between $(P_t^s)_{**}X$ and $(P_t^s)_{**}DX$, as in [**Mar83**, Proposition 19.12], but we do not need it.

We end this section with a note on operations on P_t^s-homology.

PROPOSITION 2.2.12. (a) *When $p = 2$,*
$$(P_t^s)_{**}P_t^s = (P_t^s)_{**}p_t^s = H_*(A \,\square_{D[\xi_t^{p^s}]}\mathbf{F}_p, P_t^s) \otimes (P_t^s)_{**}.$$
(b) *When p is odd,*
$$(Q_n)_{**}Q_n = (Q_n)_{**}q_n = H_*(A \,\square_{E[\tau_n]}\mathbf{F}_p, Q_n) \otimes (Q_n)_{**}.$$
(c) *When p is odd,*
$$(P_t^s)_{**}P_t^s = (P_t^s)_{**}p_t^s.$$

PROOF. To fix notation, we do part (a); the other cases are virtually identical. By Lemma 2.2.4, $P_t^s = b_{ts}^{-1}p_t^s$, so
$$P_t^s \wedge P_t^s = b_{ts}^{-1}p_t^s \wedge b_{ts}^{-1}p_t^s = b_{ts}^{-1}p_t^s \wedge p_t^s = P_t^s \wedge p_t^s.$$
This proves the first equality. The rest follows from Corollary 2.2.5 and Remark 2.2.6. □

Margolis has calculated $H_*(A \,\square_{D[\xi_t^{p^s}]}\mathbf{F}_p, P_t^s)$ for $s = 0$ at the prime 2 [**Mar83**, Corollary 19.26]; similar calculations work at odd primes for $H_*(A \,\square_{E[\tau_n]}\mathbf{F}_p, Q_n)$. We give details in Proposition 4.6.1.

2.3. Vanishing lines for homotopy groups

In this section we prove several theorems relating the vanishing of P_t^s- and Q_n-homology groups of an object X to the homotopy groups of X; these are versions in the cochain complex category of well-known theorems about modules over the Steenrod algebra. These results are at the heart of many of the other results of the book.

Here are our main theorems. In the setting of modules over the Steenrod algebra, the first is due to Anderson-Davis [**AD73**] ($p = 2$) and Miller-Wilkerson [**MW81**] (p odd), and the second to Adams-Margolis [**AM71**] ($p = 2$) and Moore-Peterson [**MP72**] (p odd). Both of the results are proved in [**MW81**]; we follow those proofs.

2.3. VANISHING LINES FOR HOMOTOPY GROUPS

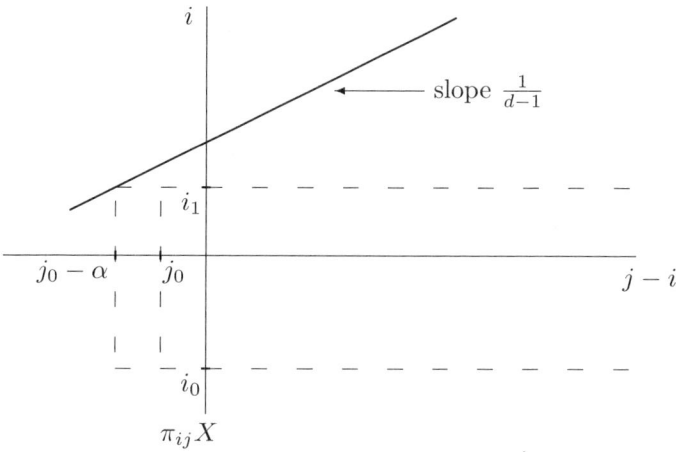

FIGURE 2.3.A. Vanishing line at the prime 2. In this picture, $\pi_{ij}X = 0$ above the slanted line (the vanishing line), to the left of the vertical line $j - i = j_0$, and below the horizontal line $i = i_0$. α is a number depending only on the slope d.

Recall that if X is an injective resolution of a comodule M, then $\pi_{**}(X) = \operatorname{Ext}_A^{**}(\mathbf{F}_p, M)$. We make heavy use of Notation 2.2.8 in this section; CL-spectra are defined in Definition 1.3.6.

THEOREM 2.3.1 (Vanishing line theorem). *Let X be a CL-spectrum. Suppose that there is a number d so that $Z(n)_{**}X = 0$ for all slopes n with $n < d$. Then $\pi_{**}X$ has a vanishing line of slope d: for some c, $\pi_{ij}X = 0$ when $j < di - c$.*

Of course in $(j - i, i)$-coordinates, this vanishing line has slope $\frac{1}{d-1}$, and the vanishing occurs above this line. See Figure 2.3.A for a picture; the numbers i_0, i_1, and j_0 there come from the definition of CL-spectra, Definition 1.3.6.

THEOREM 2.3.2. *Let X be a CL-spectrum, and assume that $Z(n)_{**}X = 0$ for all slopes n; then $\pi_{**}X$ has a "horizontal" vanishing line: for some c, $\pi_{ij}X = 0$ when $i > c$.*

COROLLARY 2.3.3. *Under the hypotheses of Theorem 2.3.2, X is in the localizing subcategory generated by $A = H\mathbf{F}_p$.*

PROOF. This follows from Lemma 1.3.8. □

For comparison purposes, we mention the following result, which we do not prove. See Definition 0.4.1 for the definition of "cofree."

THEOREM 2.3.4 ([**AM71**], [**MP72**]). *Suppose that M is a bounded below A-comodule and X is its injective resolution. Then $Z(n)_{**}X = 0$ for all slopes n if and only if M is cofree as an A-comodule.*

REMARK 2.3.5.   (a) These theorems are false without the assumption that X is CL. For example, the $(0,0)$-connective spectrum

$$X = \bigvee_{i \geq 0,\ j \geq 0} \Sigma^{i, i+j} H\mathbf{F}_p$$

is $Z(n)$-acyclic for all n, since $Z(n)_{**}H\mathbf{F}_p = 0$, but its homotopy $\pi_{**}X$ does not have a horizontal vanishing line.

(b) When we speak of phenomena "above" a vanishing line, we mean "above" in the grading of Figure 2.3.A—i.e., points $(i, j-i)$ so that $i > \frac{1}{d-1}(j-i) + b$ for some b.

(c) Suppose that X is CL with associated constants i_0, i_1, and j_0. We will see in the proof that the "intercept" c of the vanishing line in Theorem 2.3.1 may be given by $c = (d-1)i_1 - j_0 + \alpha$, where $\alpha = \alpha(d)$ is a number depending only on d. In the grading of Figure 2.3.A, $\pi_{ij}X = 0$ above the line of slope $\frac{1}{d-1}$ through the point $(j_0 - \alpha, i_1)$. To compute α, we find the smallest integer n so that the map $A \to A(n)$ is an isomorphism through degree d (i.e., so that $2^{n+1} > d$ when $p = 2$, or $2(p-1)p^n > d$ when p is odd); then

$$\alpha = \begin{cases} \displaystyle\sum_{\substack{s+t \leq n+1 \\ |\xi_t^{p^s}| \leq d}} |\xi_t^{p^s}|, & \text{if } p = 2, \\ \displaystyle\sum_{\substack{s+t \leq n \\ |\xi_t^{p^s}| \leq d}} (d + (p-1)|\xi_t^{p^s}|) + \sum_{\substack{i \leq n \\ |\tau_i| \leq d}} |\tau_i|, & \text{if } p \text{ is odd.} \end{cases}$$

For Theorem 2.3.2, the intercept is i_1—the homotopy $\pi_{ij}X$ is zero if $i > i_1$.

(d) For both of these theorems (in fact, for all of the results of this section), we can weaken condition (b) of Definition 1.3.6 slightly—it suffices to assume that for all i, there is a $j_0 = j_0(i)$ so that $\pi_{ij}X = 0$ if $j - i < j_0$. With this assumption, one replaces j_0 in the above formulas for c with $\min\{j_0(i) \mid i_0 \leq i \leq i_1\}$.

One can generalize these results somewhat. The same proofs carry over essentially unchanged, so to keep the notation simple we prove Theorems 2.3.1 and 2.3.2, rather than the following.

THEOREM 2.3.6. *Let X be a CL-spectrum, and fix a quotient Hopf algebra A' of A.*

(a) *Suppose that there is a number d so that $Z(n)_{**}X = 0$ for all slopes n with $n < d$ and $y_n \neq 0$ in A'. Then $HA'_{**}X$ has a vanishing line of slope d: for some c, $HA'_{ij}X = 0$ when $j < di - c$.*

(b) *If $Z(n)_{**}X = 0$ for all slopes n with $y_n \neq 0$ in A', then $HA'_{**}X$ has a "horizontal" vanishing line; hence $HA' \wedge X$ is in the localizing subcategory generated by $A = H\mathbf{F}_p$.*

(c) *Suppose that X is an injective resolution of a bounded below A-comodule M. Then $Z(n)_{**}X = 0$ for all slopes n with $y_n \neq 0$ in A' if and only if M is a cofree A'-comodule.*

Of course, HA'_{**} relates to $\text{Ext}_{A'}$ the same way π_{**} relates to Ext_A, so one should view this result as a vanishing line theorem over A'. (One could also work in $\mathsf{Stable}(A')$ and prove a result about π_{**} in that category, but it seems better to work in $\mathsf{Stable}(A)$ whenever possible.)

2.3.1. Proof of Theorems 2.3.1 and 2.3.2 when $p = 2$. We give the proofs of Theorems 2.3.1 and 2.3.2 when $p = 2$. In the next subsection we indicate the changes necessary when working at odd primes.

2.3. VANISHING LINES FOR HOMOTOPY GROUPS

The proof has two parts: in Lemma 2.3.11, we show that it suffices to check that $HB_{**}X$ has a vanishing line for all finite-dimensional quotient Hopf algebras B of A. In the rest of the lemmas, we verify the presence of the vanishing line for finite-dimensional quotients by induction on dimension. We give the full proofs of the theorems at the end of the subsection.

As noted in Remark 2.1.3, the classification theorem 2.1.1 of quotient Hopf algebras B of A tells us that if B is finite-dimensional, then for some s and t we have the following conditions:

- $\xi_t^{2^s} \neq 0$ in B,
- $\xi_t^{2^{s+1}} = 0$ in B,
- $\xi_j^{2^s} = 0$ in B for all $j < t$.

Hence by Lemma 2.1.2, there is a Hopf algebra extension

$$E[\xi_t^{2^s}] \to B \to C,$$

and so by Lemma 1.2.15 an element $h_{ts} \in HB_{1,|\xi_t^{2^s}|}$ and a cofibration

$$\Sigma^{1,|\xi_t^{2^s}|} HB \xrightarrow{h_{ts}} HB \to HC.$$

Given this situation, we have the following result. (Given a spectrum Y and a self-map $v: \Sigma^{ij}Y \to Y$, we say that v *acts nilpotently* on $\pi_{**}Y$ if for all $y \in \pi_{**}Y$, some power of $\pi_{**}(v)$ annihilates y.)

LEMMA 2.3.7. *Assume that B is a finite-dimensional quotient Hopf algebra of A, and fix s and t as above.*

(a) *If $s \geq t$, then $h_{ts} \in HB_{**}$ is nilpotent; hence the self-map $h_{ts}: HB \to HB$ is nilpotent.*

(b) *Given an object X in $\mathsf{Stable}(A)$, consider the cofibration*

$$\Sigma^{1,|\xi_t^{2^s}|} HB \wedge X \xrightarrow{h_{ts} \wedge 1_X} HB \wedge X \to HC \wedge X.$$

If $h_{ts} \wedge 1_X$ acts nilpotently on $HB_{**}X$, and if $HC_{**}X$ has a vanishing line of slope d, then $HB \wedge X$ has a vanishing line of slope d. The difference in intercepts depends only on d and $|\xi_t^{2^s}|$—it is independent of X.

REMARK 2.3.8. In fact, the nilpotence of h_{ts} in part (a) does not depend on B being finite-dimensional. See Theorem B.2.1 for a generalization, due to Lin.

PROOF. Lin proved part (a) in [**Lin77a**, Corollary 3.2]; see also Anderson-Davis [**AD73**] and Miller-Wilkerson [**MW81**, Proposition 4.1].

For part (b), we look at the long exact sequence in homotopy coming from the given cofibration (we write h for $\pi_{**}(h_{ts} \wedge 1_X)$):

$$\cdots \to HC_{i-1,j}X \to HB_{i-1,j-|\xi_t^{2^s}|}X \xrightarrow{h} HB_{ij}X \to HC_{ij}X \to \cdots.$$

Fix a bidegree (i,j) above the vanishing line for $HC_{**}X$. There are two cases: suppose first that $|\xi_t^{2^s}| > d$. The exact sequence tells us that

$$HB_{i-1,j-|\xi_t^{2^s}|}X \xrightarrow{h} HB_{ij}X$$

is an epimorphism. Since $|\xi_t^{2^s}| > d$, then $(i-1, j-|\xi_t^{2^s}|)$ is above the vanishing line for $HC_{**}X$. Hence

$$HB_{i-k,j-k|\xi_t^{2^s}|}X \xrightarrow{h} HB_{i-(k-1),j-(k-1)|\xi_t^{2^s}|}X$$

is an epimorphism for all $k > 0$. Also since $|\xi_t^{2^s}| > d$, then one can see that for $k \gg 0$, $(i-k, j-(k-1)|\xi_t^{2^s}|)$ is a bidegree above the vanishing line for $HC_{**}X$, so

$$HB_{i-k, j-k|\xi_t^{2^s}|}X \xrightarrow{h} HB_{i-(k-1), j-(k-1)|\xi_t^2|}X$$

is an isomorphism. Since h acts nilpotently on $HB_{**}X$, though, these groups must be zero for k large; since they surject onto $HB_{ij}X$, then $HB_{ij}X = 0$. Note that in this case, the vanishing line for $HC_{**}X$ is the same as that for $HB_{**}X$.

Suppose, on the other hand, that $|\xi_t^{2^s}| \leq d$. If (i,j) is a bidegree above the vanishing line, then so is $(i+1, j)$, so the map

$$HB_{i, j-|\xi_t^{2^s}|}X \xrightarrow{h} HB_{i+1, j}X$$

is an isomorphism. Arguing as above, we see that the group $HB_{i, j-|\xi_t^{2^s}|}X$ must be zero. In this case, the intercept of the HB vanishing line changes by $|\xi_t^{2^s}|$: $HB_{ij}X = 0$ when $j < di - c - |\xi_t^{2^s}|$. □

LEMMA 2.3.9. *Fix a spectrum X satisfying condition* (a) *of Definition 1.3.6.*

(a) *Given an extension*

$$E[\xi_t^{2^s}] \to B \to C$$

where B and C are quotient Hopf algebras of A, if $|\xi_t^{2^s}| \geq d$ and if $HC_{**}X$ has a vanishing line of slope d, then $HB_{**}X$ has a vanishing line of slope d. In fact, $HB_{**}X$ has the same vanishing line as $HC_{**}X$.

(b) *Given finite-dimensional quotient Hopf algebras $B \twoheadrightarrow C$ of A, if $HC_{**}X$ has a vanishing line of slope d and if the map $B \twoheadrightarrow C$ is an isomorphism in dimensions less than d, then $HB_{**}X$ has a vanishing line of slope d. In fact, $HB_{**}X$ has the same vanishing line as $HC_{**}X$.*

PROOF. Part (a): (This proof is based on that of [**MW81**, Proposition 3.2].) We assume that $HC_{i,j}X = 0$ when $j < di - c$, and we want to show that $HB_{i,j}X = 0$ when $j < di - c$. We prove this by induction on i. By Lemma 1.2.15, we have a cofibration

$$\Sigma^{1, |\xi_t^{2^s}|} HB \wedge X \xrightarrow{h_{ts}} HB \wedge X \to HC \wedge X.$$

This gives us a long exact sequence in homotopy:

(2.3.10) $\quad \cdots \to HB_{i-1, j-|\xi_t^{2^s}|}X \to HB_{i,j}X \to HC_{i,j}X \to \cdots .$

Since X satisfies condition (a) of Definition 1.3.6, then for i sufficiently small and for all j, we have

$$\pi_{i,j}(X) = 0 = HB_{i,j}X = HC_{i,j}X.$$

This starts the induction: if i_0 is the smallest value of i for which $HB_{i,j}X$ is nonzero, then we have an inclusion $HB_{i_0, j}X \hookrightarrow HC_{i_0, j}X$ (for all j).

The inductive step is also easy: if we have i and j with $j < di - c$, then $j - |\xi_t^{2^s}| < d(i-1) - c$; hence both $HC_{i,j}X$ and $HB_{i-i, j-|\xi_t^{2^s}|}X$ are zero. So we apply exactness in the long exact sequence (2.3.10).

Part (b): Given $B \twoheadrightarrow C$ with B finite-dimensional (or more generally, with C of finite index in B), then there is a sequence of extensions

$$E[\xi_{t_1}^{2^{s_1}}] \to B \to B_1,$$
$$E[\xi_{t_2}^{2^{s_2}}] \to B_1 \to B_2,$$
$$\vdots$$
$$E[\xi_{t_n}^{2^{s_n}}] \to B_{n-1} \to C.$$

(See [**HS98**, Lemma A.15] or [**MW81**, Lemmas 3.4–3.5], for example.) If, furthermore, $B \twoheadrightarrow C$ is an isomorphism through degree $d-1$, then each $\xi_{t_i}^{2^{s_i}}$ has degree at least d. So apply induction and part (a). $\square$

Lemma 2.3.9 is a special case of the following; we have stated and proved them separately because at odd primes we need a variant of the above result, while the following holds as stated.

LEMMA 2.3.11. *Fix an object X satisfying condition* (a) *of Definition* 1.3.6. *Given a surjection of coalgebras $f: A \twoheadrightarrow B$, if f is an isomorphism below degree d, and if $HB_{**}X$ has a vanishing line of slope d, then $HA_{**}X$ has a vanishing line of slope d. In fact, they have the same vanishing line.*

PROOF. We have an injection of comodules $\mathbf{F}_2 = A \square_A \mathbf{F}_2 \to A \square_B \mathbf{F}_2$, and the cokernel is zero below dimension d. Taking injective resolutions gives a cofibration

$$\Sigma^{1,0} Z \to HA \to HB \to Z,$$

where Z is $(0, d)$-connective (Definition 1.3.4). So to show that $HA_{ij}X = \pi_{ij}X = 0$ if $j < di - c$, one argues by induction on i just as in Lemma 2.3.9(a). $\square$

LEMMA 2.3.12. *Let X be a spectrum satisfying conditions* (b) *and* (c) *of Definition* 1.3.6. *Suppose that E is a finite-dimensional elementary quotient Hopf algebra of A. If $(P_t^s)_{**}X = 0$ whenever $\xi_t^{2^s} \neq 0$ in E, then $HE_{**}X$ has a horizontal vanishing line. In fact, given i_1 as in condition* (c), *then $HE_{i*}X = 0$ if $i > i_1$.*

PROOF. Since the dual of E is an exterior algebra, we abuse notation and write $E = E[x_1, x_2, \ldots, x_n]$ where each x_k is primitive; in other words, each x_k is equal to $\xi_t^{2^s}$ for some s and t. We point out that the x_k's have distinct degrees.

Now HE_{**} is a polynomial algebra on n generators; we write $v_k \in HE_{1,|x_k|}$ for the generator corresponding to x_k. We also use v_k to denote the corresponding self-map of HE, and we let $HE/(v_k)$ be its cofiber. Note that the self-map v_k induces multiplication by v_k on homotopy, and since $v_k \in HE_{**}$ is a nonzero divisor,

$$\pi_{**}(HE/(v_k)) = (\pi_{**}HE)/(v_k).$$

We define $HE/(v_{k_1}, \ldots, v_{k_m})$ similarly, assuming that the numbers $k_1, \ldots, k_m$ are distinct. Hence if ℓ is an integer with $1 \leq \ell \leq n$ and $\ell \notin \{k_1, \ldots, k_m\}$, then $v_\ell \colon HE \to HE$ induces a self-map of $HE/(v_{k_1}, \ldots, v_{k_m})$.

Also note that if $x_k = \xi_t^{2^s}$, then by a change-of-coalgebras, we have

$$(P_t^s)_{**}X = v_k^{-1}(HE/(v_1, \ldots, \hat{v}_k, \ldots, v_n))_{**}X.$$

We refer to this as the x_k-*homology* of X.

We claim for any set $\{k_1, \ldots, k_m\} \subseteq \{1, \ldots, n\}$,

$$(HE/(v_{k_1}, \ldots, v_{k_m}))_{**}(X)$$

has a horizontal vanishing line. We prove this by induction on $n - m$. (We will have proved the lemma when $m = 0$, i.e., when $n - m = n$.)

We have to deal with the first few cases before we can apply the inductive step. If $n - m = 0$, then $\{k_1, \ldots, k_m\} = \{1, \ldots, n\}$, so $HE/(v_{k_1}, \ldots, v_{k_m}) = H\mathbf{F}_2$, and by assumption, $H\mathbf{F}_{2**}X$ has a horizontal vanishing line with intercept i_1. If $n - m = 1$, then we let k be the integer so that

$$\{k\} \cup \{k_1, \ldots, k_m\} = \{1, 2, \ldots, n\}.$$

To simplify the notation, we let $HE_m = HE/(v_{k_1}, \ldots, v_{k_m})$. Then we have a cofibration

$$HE_m \wedge X \xrightarrow{v_k} HE_m \wedge X \to H\mathbf{F}_2 \wedge X.$$

By hypothesis, $H\mathbf{F}_{2**}X$ has a horizontal vanishing line, so the map v_k induces an isomorphism in π_{ij} for $i > i_1$. On the other hand, the x_k-homology of X is zero, and the x_k-homology of X is equal to $v_k^{-1}(HE_m)_{**}X$. Since v_k induces an isomorphism for $i > i_1$, and since this localization is zero, then v_k must be zero when $i > i_1$. So $(HE_m)_{**}X$ has a horizontal vanishing line with intercept i_1.

Now fix m with $n - m \geq 2$, and assume that

$$(HE/(v_{\ell_1}, \ldots, v_{\ell_t}))_{**}X$$

has a horizontal vanishing line whenever $n - t < n - m$. As above, let $HE_m = HE/(v_{k_1}, \ldots, v_{k_m})$. Pick distinct integers $k, \ell \leq n$ which are not in $\{k_1, \ldots, k_m\}$, and consider the following diagram, in which each row and column is a cofibration:

$$\begin{array}{ccccc}
\Sigma^{2,|x_k|+|x_\ell|}HE_m & \xrightarrow{v_k} & \Sigma^{1,|x_\ell|}HE_m & \longrightarrow & \Sigma^{1,|x_\ell|}HE_m/(v_k) \\
{\scriptstyle v_\ell}\downarrow & & {\scriptstyle v_\ell}\downarrow & & \downarrow \\
\Sigma^{1,|x_k|}HE_m & \xrightarrow{v_k} & HE_m & \longrightarrow & HE_m/(v_k) \\
\downarrow & & \downarrow & & \downarrow \\
\Sigma^{1,|x_k|}HE_m/(v_\ell) & \longrightarrow & HE_m/(v_\ell) & \longrightarrow & HE_m/(v_k, v_\ell).
\end{array}$$

Now smash this diagram with X. By induction, all of the spectra $HE_m/(v_k, v_\ell) \wedge X$, $HE_m/(v_k) \wedge X$, and $HE_m/(v_\ell) \wedge X$ have horizontal vanishing lines with intercept i_1. Now we apply $\pi_{ij}(-)$; for $i > i_1$, the maps labeled v_k and v_ℓ induce isomorphisms on π_{ij}, so we have

$$\pi_{i,j}HE_m \wedge X \xrightarrow[\cong]{v_\ell^{-1} \circ v_k} \pi_{i,j+|x_k|-|x_\ell|}HE_m \wedge X.$$

This isomorphism, combined with the facts that $|x_k| \neq |x_\ell|$ and that $\pi_{ij}(HE_m \wedge X)$ is zero when $j \ll 0$, implies that $\pi_{ij}(HE_m \wedge X) = 0$ for all j; i.e., $\pi_{**}(HE_m \wedge X)$ has the predicted horizontal vanishing line. This completes the inductive step, and hence the proof. $\square$

PROOF OF THEOREM 2.3.1. By Lemma 2.3.11, if we know that $HB_{**}X$ has a vanishing line of slope d for all finite-dimensional quotient Hopf algebras B of A, then $\pi_{**}(X)$ will, also. (For example, the map $A \to A(n)$ is an isomorphism through degree $2^{n+1} - 1$, so apply the lemma to the case $B = A(n)$ where n is chosen large enough so that $2^{n+1} - 1 \geq d$.)

Now we show that $HB_{**}X$ has a vanishing line of slope d for all quotients B of A with $\dim_{\mathbf{F}_2} B < \infty$, by induction on $\dim_{\mathbf{F}_2} B$: The case where $\dim_{\mathbf{F}_2} B = 1$ (i.e.,

$B = \mathbf{F}_2$) is taken care of by condition (c) of the definition of CL, Definition 1.3.6, so we move on to the inductive step. By Remark 2.1.3, there is a Hopf algebra extension
$$E[\xi_t^{2^s}] \to B \to C.$$
By hypothesis, $HC_{**}X$ has a vanishing line of slope d, and we want to produce a vanishing line for $HB_{**}X$. There are several cases:

(1) If $s \geq t$, then we are done by Lemma 2.3.7.

(2) If $s < t$ and $|\xi_t^{2^s}| > d$, then we are done by Lemma 2.3.9(a).

(3) If $s < t$ and $|\xi_t^{2^s}| \leq d$, then we have to prove something. We may assume that $\xi_j = 0$ in B for $j < t$. Let
$$B = \mathbf{F}_2[\xi_t, \xi_{t+1}, \xi_{t+2}, \ldots]/(\xi_t^{2^{s+1}}, \xi_{t+1}^{2^{n_1}}, \xi_{t+2}^{2^{n_2}}, \xi_{t+3}^{2^{n_3}}, \ldots),$$
$$E = \mathbf{F}_2[\xi_t, \xi_{t+1}, \xi_{t+2}, \ldots]/(\xi_t^{2^{s+1}}, \xi_{t+1}^{2^{m_1}}, \xi_{t+2}^{2^{m_2}}, \xi_{t+3}^{2^{m_3}}, \ldots),$$
where $m_i = \max(\min(n_i, s-i), 0)$. By Proposition 2.1.7, E is an elementary quotient Hopf algebra of A (and of B). Furthermore, if $\xi_v^{2^u} \neq 0$ in E, then $|\xi_v^{2^u}| \leq |\xi_t^{2^s}|$. By hypothesis on X, then, $(P_v^u)_{**}X = 0$ when $\xi_v^{2^u} \neq 0$ in E. So we apply Lemma 2.3.12 to conclude that $HE_{**}X$ has a horizontal vanishing line with intercept i_1. Since X satisfies condition (b) of Definition 1.3.6, then $HE_{**}X$ also has a vanishing line of slope $|\xi_t^{2^s}| + 1$ (the line with this slope through the point (i_1, j_0)). We then apply Lemma 2.3.9(b) to the case $B \twoheadrightarrow E$ to conclude that $HB_{**}X$ has a vanishing line of slope $|\xi_t^{2^s}| + 1$ (and in fact the same vanishing line).

Now, the map h_{ts} acts at slope $|\xi_t^{2^s}|$, and hence acts nilpotently on $HB_{**}X$. By Lemma 2.3.7(b), then, the vanishing line for $HC_{**}X$, which has slope d, gives one for $HB_{**}X$. □

PROOF OF THEOREM 2.3.2. By Proposition 2.1.5, it suffices to show that there is a uniform horizontal vanishing line for all of the groups $HA(n)_{**}X$. Lemma 2.3.12 and Lemma 2.3.9 together show that each $HA(n)_{**}X$ has a horizontal vanishing line, and in fact these results identify the vanishing line: $HA(n)_{i*}X = 0$ if $i > i_1$, with i_1 as in condition (c) of Definition 1.3.6. □

2.3.2. Changes necessary when p is odd. The proofs of Theorems 2.3.1 and 2.3.2 given above go through with a few changes when the prime p is odd; we indicate those changes in this subsection.

We start with the same set-up as when $p = 2$: we have an extension of Hopf algebras in one of the following forms:

(E) $\quad E[\tau_n] \to B \to C,$

(D) $\quad D[\xi_t^{p^s}] \to B \to C.$

If case (E) arises, then Lemma 2.3.7(b) carries over as stated (writing v_n for the homotopy element associated to the element τ_n, rather than h_{ts}).

Otherwise extension (D) arises, and we may assume that we have

- $\xi_t^{p^s} \neq 0$ in B,
- $\xi_t^{p^{s+1}} = 0$ in B,
- $\xi_j^{p^s} = 0$ in B for all $j < t$.

Lemma 1.2.15 then gives an element $b_{ts} \in HB_{2,p|\xi_t^{p^s}|}$, and hence a cofibration

(2.3.13) $\qquad \Sigma^{2,p|\xi_t^{p^s}|}HB \xrightarrow{b_{ts}} HB \to \widetilde{HC} \to \Sigma^{1,p|\xi_t^{p^s}|}HB,$

where $\widetilde{HC}$ is the cofiber of a self-map of HC, as in Lemma 1.2.15. Hence if $HC_{**}X$ has a vanishing line of slope d, then so does $\widetilde{HC}_{**}X$. More precisely, if $HC_{ij}X = 0$ when $j < di - c$, then $\widetilde{HC}_{ij}X = 0$ when $j < di - c + \min(0, |\xi_t^{p^s}| - d)$.

Here is the odd prime analogue of Lemma 2.3.7.

LEMMA 2.3.14. (a) *Given s and t as above so that we have extension* (D), *if $s \geq t$, then $b_{ts} \in HB_{**}$ is nilpotent; hence the self-map $b_{ts} \colon HB \to HB$ is nilpotent.*

(b) *Given extension* (D) *and an object X, consider the cofibration*

$$\Sigma^{2,p|\xi_t^{p^s}|}HB \wedge X \xrightarrow{b_{ts} \wedge 1_X} HB \wedge X \to \widetilde{HC} \wedge X.$$

If $b_{ts} \wedge 1_X$ acts nilpotently on $HB_{**}X$, and if $\widetilde{HC}_{**}X$ has a vanishing line of slope d, then $HB \wedge X$ has a vanishing line of slope d. The difference in intercepts depends only on d and $|\xi_t^{p^s}|$—it is independent of X.

PROOF. For part (a), see [**MW81**, Proposition 4.1].
Part (b) is proved just as in the $p = 2$ case. $\square$

The odd prime version of Lemma 2.3.9 is as follows.

LEMMA 2.3.15. *Fix an object X satisfying condition* (a) *of Definition* 1.3.6.

(a) *Given an extension*

$$D[\xi_t^{p^s}] \to B \to C,$$

if $\frac{p|\xi_t^{2s}|}{2} \geq d$ and if $HC_{**}X$ has a vanishing line of slope d, then $HB_{**}X$ has a vanishing line of slope d. In fact, $HB_{**}X$ has the same vanishing line as $\widetilde{HC}_{**}X$.

(b) *Given an extension*

$$E[\tau_n] \to B \to C,$$

if $|\tau_n| \geq d$ and if $HC_{**}X$ has a vanishing line of slope d, then $HB_{**}X$ has a vanishing line of slope d. In fact, $HB_{**}X$ has the same vanishing line as $HC_{**}X$.

(c) *Given finite-dimensional quotients $B \twoheadrightarrow C$ of A, if $HC_{**}X$ has a vanishing line of slope d and if the map $B \twoheadrightarrow C$ is an isomorphism in odd dimensions less than d and in even dimensions less than $\frac{2d}{p}$, then $HB_{**}X$ has a vanishing line of slope d. Furthermore, the difference in intercept between the two vanishing lines is independent of X.*

PROOF. Part (a) is proved just as is Lemma 2.3.9(a), but based on the cofibration (2.3.13). Part (b) is the same as Lemma 2.3.9(a), except for the characteristic of the ground field, which is not relevant. Part (c) is proved, as in Lemma 2.3.9(b), by induction and parts (a) and (b). Since in (a), the vanishing line for $\widetilde{HC}_{**}X$ may have a different intercept than that for $HC_{**}X$, the intercept for $HB_{**}X$ will change as the induction proceeds, but it will change by amounts dependent only on d. $\square$

Lemma 2.3.11 holds as stated, regardless of the prime involved.
Here is the analogue of Lemma 2.3.12.

LEMMA 2.3.16. *Let X be an object satisfying conditions* (b) *and* (c) *of Definition 1.3.6. Suppose that E is a finite-dimensional elementary quotient Hopf algebra of A. If $(P_t^s)_{**}X = 0$ whenever $\xi_t^{2^s} \neq 0$ in E, and $(Q_n)_{**}X = 0$ whenever $\tau_n \neq 0$ in E, then $HE_{**}X$ has a horizontal vanishing line. In fact, given i_1 as in condition* (c), *then $HE_{i*}X = 0$ if $i > i_1$.*

The proof is the same as that for Lemma 2.3.12, using cofibrations of the form (2.3.13) repeatedly. We also need to point out that the slopes $\frac{p|\xi_t^{p^s}|}{2}$ and $|\tau_n|$ are all distinct. (Note also that in this case, we may choose d as large as we like, so that a horizontal vanishing line for $HC_{**}X$ induces the same one for $\widetilde{HC}_{**}X$.)

Finally, given these modified tools, the proofs of the main theorems go through essentially unchanged.

2.4. Self-maps via vanishing lines

In this section we use vanishing lines to construct self-maps of finite spectra. For each finite X, we construct one self-map; we give many others in Theorem 6.1.3. We make heavy use of Notation 2.2.8 in this section.

Our approach for this section is based on some recent work in ordinary stable homotopy theory, as described in work of Ravenel [**Rav86**] and Hopkins and Smith [**HS98**]. For now, we view the objects P_t^s and Q_n (i.e., the $Z(n)$'s) as the analogues of Morava K-theories. This is not a perfect analogy, because the Morava K-theories detect nilpotence, while the $Z(n)$'s do not.

The following definition is based on the definition of "v_n-map" in ordinary stable homotopy theory; see [**HS98**] and [**Rav92**]. We will investigate and generalize this definition in Section 6.2.

NOTATION 2.4.1. Fix a slope n. The ring $z(n)_{**}$ is either polynomial on one generator or has such a polynomial algebra as a tensor factor; we call the polynomial generator u_n. (In the notation of Remark 2.1.3 and Proposition 2.2.3, u_n is one of h_{ts}, b_{ts}, or v_n.) We write $\eta_{Z(n)}$ for the composite $S^0 \xrightarrow{\eta} z(n) \to Z(n)$, where η is the "unit map" of the spectrum $z(n)$—i.e., the map given by Proposition 1.2.8(b).

DEFINITION 2.4.2. Fix a spectrum X and a slope n.

(a) A self-map $f \in [X, X]_{**}$ is a u_n-*map* if for some j,
$$\eta_{Z(n)} \wedge f = u_n^j \wedge 1_X$$
as elements of $[X, Z(n) \wedge X]_{**}$.

(b) We say that X is of *type n* if $Z(d)_{**}X = 0$ for $d < n$, and $Z(n)_{**}X \neq 0$.

Notice that if $Z(n)_{**}X = 0$, then the zero map of X is a u_n-map.

We point out that in the ordinary stable homotopy category, Ravenel showed in [**Rav84**, Theorem 2.11] that for any finite spectrum X, $K(n)_*X \neq 0 \Rightarrow K(n+1)_*X \neq 0$. The analogous statement here, with $Z(n)_{**}$ rather than $K(n)_*$, does not hold. See Proposition 6.8.2 (and also [**Pal96b**, Proposition 3.10 and Theorem A.1]) for the correct statement when $p = 2$, and for a guess in the odd prime case. We do know that if X is a nonzero finite spectrum, then by Theorem 2.3.2 and Corollary 6.6.8, for some n we have $Z(n)_{**}X \neq 0$. (One could also use the Atiyah-Hirzebruch spectral sequence to show this.)

The following first appeared for modules in [**Pal92**]. It is a slight generalization of a result of Hopkins and Smith [**HS98**]. See Theorems 5.1.2 and 6.1.3 for stronger results when $p = 2$.

THEOREM 2.4.3. *Fix a finite spectrum X of type n. Then for some k, there is a non-nilpotent u_n-map*
$$v\colon \Sigma^{k,kn}X \to X.$$

To prove this, we need a "relative vanishing line" result; this is a generalization of a standard result—see [**Rav86**, Lemma 3.4.9], for instance.

LEMMA 2.4.4. *Fix a slope n. Suppose that X is a CL-spectrum, and suppose that X is of type at least n; hence $\pi_{**}X$ has a vanishing line of slope n. Given $m \geq 0$, let $M = M(m)$ be the number below which degree $A \twoheadrightarrow A(m)$ is an isomorphism. Then the Hurewicz map $h\colon \pi_{**}X \to HA(m)_{**}X$ is an isomorphism above a line of slope n: for some c independent of m, h is an isomorphism on π_{ij} when $j < ni + M - c$.*

PROOF. Let W denote the fiber of the map $S^0 \to HA(m)$. Since the kernel of the map $A \to A(m)$ is zero below degree M, if we choose numbers i_0, i_1, and j_0 so that X satisfies Definition 1.3.6, then $W \wedge X$ satisfies the conditions with the numbers i_0, i_1, and $j_0 + M$. Hence for some number c, the vanishing line theorem 2.3.1 implies that $\pi_{ij}(W \wedge X) = 0$ when $j < ni + M - c$. By Remark 2.3.5, c depends only on n, i_1, and j_0. □

We have the following result based on work of Lin [**Lin77a**] and Wilkerson [**Wil81**]; this follows immediately from [**HS98**, Theorem 4.13], stated above as Theorem 0.4.10.

PROPOSITION 2.4.5. *Fix m and consider the quotient Hopf algebra $A(m)$ of A.*

(a) *Let $p = 2$. Fix integers $s < t$ with $\xi_t^{2^s}$ nonzero in $A(m)$ (i.e., with $s + t \leq m + 1$). Then the restriction map*
$$\mathrm{Ext}^{**}_{A(m)}(\mathbf{F}_2, \mathbf{F}_2) \to \mathrm{Ext}^{**}_{\mathbf{F}_2[\xi_t]/(\xi_t^{2^{s+1}})}(\mathbf{F}_2, \mathbf{F}_2) \cong \mathbf{F}_2[h_{t0}, h_{t1}, \ldots, h_{ts}]$$
is "surjective modulo nilpotence" (i.e., for every element in the codomain, some power of it is in the image). Hence for some $i = i(m)$, there is a non-nilpotent element
$$w \in \mathrm{Ext}^{i,i|\xi_t^{2^s}|}_{A(m)}(\mathbf{F}_2, \mathbf{F}_2) = HA(m)_{i,i|\xi_t^{2^s}|}$$
which restricts to h_{ts}^i.

(b) *Let p be odd. Fix integers $s < t$ with $\xi_t^{p^s}$ nonzero in $A(m)$ (i.e., with $s + t \leq m$). Then the restriction map*
$$\mathrm{Ext}^{**}_{A(m)}(\mathbf{F}_p, \mathbf{F}_p) \longrightarrow \mathrm{Ext}^{**}_{\mathbf{F}_p[\xi_t]/(\xi_t^{p^{s+1}})}(\mathbf{F}_p, \mathbf{F}_p)$$
$$\parallel$$
$$\mathbf{F}_p[b_{t0}, \ldots, b_{ts}] \otimes \Lambda[h_{t0}, \ldots, h_{ts}]$$
is surjective modulo nilpotence. Hence for some $j = j(m)$, there is a non-nilpotent element
$$w \in \mathrm{Ext}^{2j, jp|\xi_t^{p^s}|}_{A(m)}(\mathbf{F}_p, \mathbf{F}_p) = HA(m)_{2j, jp|\xi_t^{p^s}|}$$
which restricts to b_{ts}^j.

(c) *Let p be odd. Fix an integer t with τ_t nonzero in $A(m)$ (i.e., with $t \leq m$). Then the restriction map*

$$\operatorname{Ext}^{**}_{A(m)}(\mathbf{F}_p, \mathbf{F}_p) \to \operatorname{Ext}^{**}_{E[\tau_t]}(\mathbf{F}_p, \mathbf{F}_p) \cong \mathbf{F}_p[v_t]$$

is surjective mod nilpotence. Hence for some $k = k(m)$, there is a non-nilpotent element

$$w \in \operatorname{Ext}^{k,k|\tau_t|}_{A(m)}(\mathbf{F}_p, \mathbf{F}_p) = HA(m)_{k,k|\tau_t|}$$

which restricts to v_t^k.

We need the following lemma. Recall from Notation 0.5.1 that $D[y]$ denotes the Hopf algebra $\mathbf{F}_p[y]/(y^p)$ with y primitive.

LEMMA 2.4.6. *Fix a slope n, and choose an integer m large enough that y_n is nonzero in $A(m)$ (and hence so that $D[y_n]$ or $E[y_n]$ is a quotient coalgebra of $A(m)$ over which $A(m)$ is injective).*

(a) *Then the map $HA(m)_{**} \to z(n)_{**}$ is an algebra map, and some power of the polynomial generator u_n of $z(n)_{**}$ is in the image.*

(b) *For any object X, the following diagram commutes:*

$$\begin{array}{ccc} [X, HA(m) \wedge X]_{**} & \longrightarrow & [X, z(n) \wedge X]_{**} \\ {\scriptstyle -\wedge X}\uparrow & & \uparrow{\scriptstyle -\wedge X} \\ HA(m)_{**} & \longrightarrow & z(n)_{**}. \end{array}$$

(c) *For any object X with $Z(n)_{**}X \neq 0$, the map $z(n)_{**} \to [X, z(n) \wedge X]_{**}$ is an injection.*

PROOF. Part (a) follows from Proposition 1.2.8(b) and Proposition 2.4.5.

Part (b) follows from the fact that $[X, z(n) \wedge X]_{**}$ is isomorphic to cochain homotopy classes of $D[y_n]$-linear (respectively, $E[y_n]$-linear) self-maps of X, and the horizontal maps are just restriction.

For part (c), we merely note that if $Z(n)_{**}X \neq 0$, then the identity map on X is not cochain homotopic to zero over $D[y_n]$ (respectively, over $E[y_n]$). □

PROOF OF THEOREM 2.4.3. Since X is of type n, so is $X \wedge DX$, where DX is the Spanier-Whitehead dual of X; hence Lemma 2.4.4 applies to $[X, X]_{**} = \pi_{**}(X \wedge DX)$. By that lemma, if we choose m large enough, then the map $[X, X]_{**} \to [X, HA(m) \wedge X]_{**}$ is an isomorphism in the bidegrees (k, kn), for all integers k. Now we apply Lemma 2.4.6 to find a non-nilpotent element $w \in HA(m)_{k,kn}$ (for some k) which maps nontrivially to $z(n)_{**}$ and to $[X, z(n) \wedge X]_{**}$. The lift of $w \wedge 1_X$ to $[X, X]_{k,kn}$ has the desired properties. □

We will need the following lemma in Section 2.5.

LEMMA 2.4.7. *Fix a finite type n spectrum X. Then for some k, the u_n-map $v \in [X, X]_{**}$ constructed in Theorem 2.4.3 is central in a band parallel to the vanishing line. This band includes the origin.*

PROOF. Consider the element $w \in HA(m)_{k,kn}$, as given by Proposition 2.4.5. Since $HA(m)_{**}$ is a commutative ring and $[X, HA(m) \wedge X]$ is an algebra over it, w maps to a central element in $[X, HA(m) \wedge X]_{**}$. By Lemma 2.4.4, $[X, X]_{**} \to [X, HA(m) \wedge X]_{**}$ is an isomorphism in a band parallel to the vanishing line; by

2.5. Construction of spectra of specified type

In this section we construct certain objects for later use; these are analogues of the "generalized Toda $V(n)$ spectra," as used by Mahowald and Sadofsky [**MS95**], among others. Some of the basic ideas are standard; most of the rest are due to them. A few of the details are different in our setting.

We start by constructing the relevant spectra. See Notations 2.2.8 and 2.4.1 for the terminology used here.

WARNING 2.5.1. Recall that u_n is the polynomial generator of $z(n)_{**}$, and its powers lie on a line of slope n; u_n is an element of either $z(n)_{1,n}$ or $z(n)_{2,2n}$, depending on the prime and whether $z(n)$ is p_t^s or q_n—see Proposition 2.2.3. Recall also that $Z(n)_{**} = u_n^{-1} z(n)_{**}$. So as to avoid dividing all of our arguments into cases, we will abuse notation and write u_n^i for the power of u_n in $Z(n)_{i,in}$ (and similarly for u_n-maps: a self-map $u_n^i: X \to X$ has bidegree (i, in)). Hence when $Z(n) = P_t^s$ and p is odd, only even powers of u_n make sense.

PROPOSITION 2.5.2. *Let p be a prime, and fix a slope n. Let $1 = d_1 < d_2 < \cdots < d_m$ be the slopes less than n. For any integers $k_1, \ldots, k_m$, there are integers $j_1, \ldots, j_m$ with $k_i \leq j_i$ for each i, so that there is a finite spectrum $F = F(u_{d_1}^{j_1}, \ldots, u_{d_m}^{j_m})$ satisfying the following.*

(a) *When $n = 1$, $F = S^0$.*
(b) *F is $Z(d)_{**}$-acyclic for $d < n$, and $Z(n)_{**} F \neq 0$.*
(c) *Hence by Theorem 2.4.3, F has a u_n-map; call it u. If $Z(n)_{**}(u)$ has bidegree $(j_{m+1}, j_{m+1} n)$, then $F(u_{d_1}^{j_1}, \ldots, u_{d_m}^{j_m}, u_n^{j_{m+1}})$ is the fiber of u. More precisely, there is a fiber sequence*

$$F(u_{d_1}^{j_1}, \ldots, u_n^{j_{m+1}}) \to F(u_{d_1}^{j_1}, \ldots, u_{d_m}^{j_m}) \xrightarrow{u_n^{j_{m+1}}} \Sigma^{-j_{m+1}, -n j_{m+1}} F(u_{d_1}^{j_1}, \ldots, u_{d_m}^{j_m}).$$

(d) *If u and v are two u_n-maps of F, then $u^i = v^j$ for some numbers i and j.*
(e) *F is self-dual, and the dual of u is a u_n-map. That is, for some numbers q and r, the Spanier-Whitehead dual DF of F is isomorphic to $\Sigma^{q,r} F$, and u maps to a u_n-map under the chain of isomorphisms*

$$[F, F] \cong [DF, DF] \cong [\Sigma^{q,r} F, \Sigma^{q,r} F] \cong [F, F].$$

PROOF. The statements of (a)-(c) indicate how the spectra are constructed. Starting with S^0, one applies Theorem 2.4.3 to find a u_1-map $u_1^{j_1}$ of it, and one lets $F(u_1^{j_1})$ be the fiber. By definition of u_1-map, $u_1^{j_1}$ induces an isomorphism on $Z(1)_{**}$, so $F(u_1^{j_1})$ is $Z(1)_{**}$-acyclic; $u_1^{j_1}$ induces zero on $Z(d)_{**}$ for $d > 1$, so $F(u_1^{j_1})$ is not $Z(d)_{**}$-acyclic for any larger value of d. One proceeds inductively. This proves (a)-(c).

Part (d): Since u and v are u_n-maps, then after raising them to suitable powers, we may assume that they agree when restricted to $[F, z(n) \wedge F]_{**}$. For any $m \geq 0$, we can raise them to large enough powers so that they agree when restricted to $[F, HA(m) \wedge F]_{**}$. But when m is large enough, this ring is isomorphic to $[F, F]_{**}$ in the bidegree of interest; hence large enough powers of our self-maps agree in $[F, F]_{**}$.

Part (e): By the definition 2.4.2 of a u_n-map, the dual of a u_n-map of a finite object X is a u_n-map of the dual DX. □

Note that the fiber sequences in part (c) give a sequence of degree zero maps
$$F = F(u_{d_1}^{j_1}, \ldots, u_{d_m}^{j_m}) \to F(u_{d_1}^{j_1}, \ldots, u_{d_{m-1}}^{j_{m-1}}) \to \cdots \to F(u_{d_1}^{j_1}) \to S^0.$$
We refer to the composite $F \to S^0$ as "projection to the top cell."

Up to suspension, the object $F(u_{d_1}^{j_1}, \ldots, u_{d_m}^{j_m})$ is the analogue in $\mathsf{Stable}(A)$ of Mahowald and Sadofsky's spectrum (in the usual stable homotopy category) $M(p^{j_0}, v_1^{j_1}, \ldots, v_{n-1}^{j_{n-1}})$. We have used the letter F rather than M since our spectra are iterated fibers rather than cofibers, as they are in [**MS95**]. Also, the letter M can get somewhat overworked; in particular, we want to avoid confusion with the functor M_n^f of Section 3.3.

Now we establish the main properties of the spectra $F(u_{d_1}^{j_1}, \ldots, u_{d_m}^{j_m})$. Thick subcategories are defined in Definition 1.3.7, as is the notation thick(Y). Bousfield classes are defined in Section 1.5.

THEOREM 2.5.3. *Fix a prime p, a slope n, and other notation as in Proposition 2.5.2, and let $F = F(u_{d_1}^{j_1}, \ldots, u_{d_m}^{j_m})$ be as in that result. Then F satisfies the following properties.*

(a) *For any finite spectrum W which is $Z(d)_{**}$-acyclic for all $d < n$, W is in thick(F).*

(b) *Hence the thick subcategory generated by F and the Bousfield class of F are both independent of the choice of exponents j_i.*

(c) *Let u denote the u_n-map of F. Suppose that ℓ_1, ..., ℓ_m are integers so that $W = F(u_{d_1}^{\ell_1}, \ldots, u_{d_m}^{\ell_m})$ exists, and let v denote the u_n-map of W as given by Proposition 2.5.2(c). Then there are integers i and j so that $u^i \wedge 1_W = 1_F \wedge v^j$ as self-maps of $F \wedge W$.*

(d) *Suppose that ℓ_1, ..., ℓ_m are integers so that $F(u_{d_1}^{\ell_1}, \ldots, u_{d_m}^{\ell_m})$ exists. If $\ell_i \gg j_i$ for each i, then there is a map*
$$F(u_{d_1}^{j_1}, \ldots, u_{d_m}^{j_m}) \to F(u_{d_1}^{\ell_1}, \ldots, u_{d_m}^{\ell_m})$$
commuting with projection to the top cell—i.e., with the map $F(\cdots) \to S^0$.

PROOF. Part (a): For each $i \leq m$, let $F(U_i) = F(u_{d_1}^{j_1}, \ldots, u_{d_i}^{j_i})$. We have cofiber sequences
$$F(U_i) \to F(U_{i-1}) \xrightarrow{u_{d_i}^{j_i}} F(U_{i-1}).$$
We claim that $1_W \wedge u_{d_i}^{j_i}$ is a nilpotent self map of $W \wedge F(U_{i-1})$. Once we know this, then Lemma 5.2.5 tells us that $W \wedge F(U_{i-1})$ is in the thick subcategory generated by $W \wedge F(U_i)$ for each i. By induction, $W \wedge F(U_0) = W$ is in the thick subcategory generated by $W \wedge F(U_m)$, which is a subcategory of the thick subcategory generated by $F(U_m) = F$.

The claim that $1_W \wedge u_{d_i}^{j_i}$ is nilpotent follows by application of the vanishing line theorem 2.3.1: the group
$$[W \wedge F(U_{i-1}), W \wedge F(U_{i-1})]_{**} = \pi_{**}(W \wedge DW \wedge F(U_{i-1}) \wedge DF(U_{i-1}))$$
has a vanishing line of slope n, and $1_W \wedge u_{d_i}^{j_i}$ acts along a line of slope $d_i < n$.

Part (b): This follows immediately from part (a).

Part (c): The proof of Proposition 2.5.2(d) works here, too. By Proposition 2.5.2(b), $F \wedge W$ is $Z(d)_{**}$-acyclic for $d < n$; hence $[F \wedge W, F \wedge W]_{**}$ has a vanishing line of slope at least n. If this slope is larger than n, then the powers of both $u \wedge 1_W$ and $1_F \wedge v$ would eventually lie above the vanishing line, and hence would both be zero. So we may assume that the vanishing line has slope equal to n. Now, as in the proof of Proposition 2.5.2(d), we know that for some i and j, the self-maps $u^i \wedge 1_W$ and $1_F \wedge v^j$ agree when restricted to

$$[F \wedge W, HA(m) \wedge F \wedge W]_{**},$$

if m is large enough. Since this ring is isomorphic to $[F \wedge W, F \wedge W]_{**}$ in the bidegree of interest, we are done.

Part (d): This follows from (c). We prove it by induction on m. When $m = 1$, then the following diagram commutes:

$$\begin{array}{ccccc} F(u_1^{j_1}) & \longrightarrow & S^0 & \xrightarrow{u_1^{j_1}} & S^0 \\ & & \| & & \downarrow u_1^{\ell_1 - j_1} \\ F(u_1^{\ell_1}) & \longrightarrow & S^0 & \xrightarrow{u_1^{\ell_1}} & S^0. \end{array}$$

Hence there is an induced map $F(u_1^{j_1}) \to F(u_1^{\ell_1})$.

Assume that we have a map

$$\begin{array}{ccc} F(u_{d_1}^{j_1}, \ldots, u_{d_{m-1}}^{j_{m-1}}) & \xrightarrow{f} & F(u_{d_1}^{\ell_1}, \ldots, u_{d_{m-1}}^{\ell_{m-1}}). \\ \| & & \| \\ F & & W \end{array}$$

We abbreviate these spectra as F and W, as indicated. Consider the following diagram, in which the rows are fiber sequences:

$$\begin{array}{ccccc} F(u_{d_m}^{j_m}) & \longrightarrow & F & \xrightarrow{u_{d_m}^{j_m}} & F \\ & & \| & & \downarrow u_{d_m}^{\ell_m - j_m} \\ F(u_{d_m}^{\ell_m}) & \longrightarrow & F & \xrightarrow{u_{d_m}^{\ell_m}} & F \\ & & \downarrow f & & \downarrow f \\ W(u_{d_m}^{\ell_m}) & \longrightarrow & W & \xrightarrow{u_{d_m}^{\ell_m}} & W. \end{array}$$

Clearly the top right square commutes. We claim that part (c) implies that the lower right square commutes, when ℓ_m is large enough. Given this, we get maps

$$F(u_{d_m}^{j_m}) \to F(u_{d_m}^{\ell_m}) \to W(u_{d_m}^{\ell_m}).$$

The composite is the desired map.

To see that the lower right square commutes, we imitate the proof of Corollary 6.2.11. The key is to use Spanier-Whitehead duality so that the question is whether $1 \wedge u_{d_m}^{\ell_m}$ and $Du_{d_m}^{\ell_m} \wedge 1$ agree as self-maps of the spectrum $DF \wedge W$. By part (c), it suffices to show that F is self-dual (up to suspension), and that the dual of $u_{d_m}^{\ell_m}$ is a u_n-map. But this is Proposition 2.5.2(e). Hence if ℓ_m is sufficiently large, the square commutes, and we are done. $\square$

We point out that since $F = F(u_{d_1}^{j_1}, \ldots, u_{d_m}^{j_m})$ is well-defined up to Bousfield class, and since its u_n-map is essentially unique, then the telescope $u_n^{-1}F$ is well-defined up to Bousfield class. But the telescope of a u_n-map of an arbitrary finite spectrum of type n could have a different Bousfield class. For example, at the prime 2, there is a non-nilpotent self-map of the sphere called d_0:

$$d_0 \colon S^{4,18} \to S^0.$$

(In the May spectral sequence, this element would be called $h_{20}^2 h_{21}^2$—see Example 6.5.7(b).) Although S^0 and S^0/d_0 are both type 0, they are not Bousfield-equivalent—S^0 sees $d_0^{-1}S^0$, while S^0/d_0 does not. Hence the telescopes $h_0^{-1}S^0$ and $h_0^{-1}(S^0/d_0)$ probably have distinct Bousfield classes.

2.6. Further discussion

As mentioned in Subsection 2.1.1, one of the main gaps in this theory for odd p is the lack of a classification of the quasi-elementary quotient Hopf algebras of A. See Appendix B.3 for a discussion of conjectures and results related to this issue.

We note that the vanishing line theorem 2.3.1 has been used many times in many papers; it provides a convenient way to get valuable information about the Adams E_2-term. Combined with newer results, such as the genericity of vanishing lines in Adams spectral sequences (Theorem 1.4.4), it is even more powerful.

Theorem 2.4.3, the result that ensures the existence of a non-nilpotent self-map of any finite object, also has been used in topological applications. For example, Hopkins and Smith used it to prove the periodicity theorem [**HS98**, Theorem 9]: they had constructed a particular spectrum X, and they used the theorem to find a v_n-map of X at the E_2-term of the Adams spectral sequence. Later, Theorem 2.4.3 was used by Sadofsky and the author in [**PS94**] to give a new proof of the periodicity theorem: we used it not only to produce a v_n-map, but also to construct the spectrum X in question. This was done by taking iterated cofibers, as in Section 2.5: when $p = 2$, the cofiber of $u_1 \colon S^0 \to S^0$ has a u_3-map (3 is the next slope after 1); the cofiber of this map has a u_6-map, etc. This was done with modules over the Steenrod algebra, and then realized at the spectrum level in the ordinary stable homotopy category.

CHAPTER 3

Chromatic structure

In this chapter we discuss "chromatic" results in Stable(A). We start in Section 3.1 by discussing Margolis' killing construction [**Mar83**, Chapter 21]. This is the analogue, in our setting, of the functor L_n^f in the ordinary stable homotopy category, so we use the same notation. We give several different constructions of the functor, and we prove various properties (e.g., for X with nice connectivity properties, if X has finite type homotopy, then so does $L_n^f X$). We also define an analogue of the functor L_n, and we show that $L_n \neq L_n^f$ if $n > 1$, at least at the prime 2.

We have been using the functor H heavily throughout this book; the homotopy groups of HB, for B a quotient of A, are the cohomology groups of the Hopf algebra B. In Section 3.2 we construct $\widehat{H}(-)$, a version of this functor whose homotopy groups are the Tate cohomology groups of B. The spectrum $\widehat{H}A(m)$ turns out to be equal to $L_n^f HA(m)$, for n sufficiently large compared to m; we use this result in Section 3.3 to prove that the "chromatic tower," the tower

$$L_0^f X \leftarrow L_1^f X \leftarrow L_2^f X \leftarrow L_3^f X \leftarrow \cdots,$$

converges to X, if X is finite. This is an extension of a theorem of Margolis [**Mar83**, Theorem 22.1].

In Section 3.4 we try to compare our chromatic filtration of $\operatorname{Ext}_A^{**}(\mathbf{F}_p, \mathbf{F}_p)$ to that of Mahowald and Shick in [**MS87**] and [**Shi88**]; we have more questions than answers, though.

In Section 3.5 we discuss some other issues related to chromatic phenomena, such as constructing chromatic towers in different orders, and relating the chromatic tower construction to the multiple complex construction of Benson and Carlson.

Warning 2.5.1 applies throughout this chapter.

3.1. Margolis' killing construction

In this section we present Margolis' killing construction. This is a (smashing) localization functor that kills off P_t^s- and Q_n-homology for $\xi_t^{p^s}$ and τ_n of large slope.

We make use of Notation 2.2.8 and 2.4.1 in this section; as remarked above, Warning 2.5.1 applies here. We defined the notions of "thick" and "localizing" subcategories in Definition 1.3.7.

DEFINITION 3.1.1. Recall from [**Mil92**] and [**HPS97**, Theorem 3.3.3] that given any thick subcategory $\mathscr{C}$ of finite spectra, there is a functor $L_{\mathscr{C}}^f$: Stable(A) → Stable(A), called *finite localization away from* $\mathscr{C}$, with these properties:

(a) $L_{\mathscr{C}}^f$ is exact. When viewed as a functor $L_{\mathscr{C}}^f$: Stable(A) → $L_{\mathscr{C}}^f$ Stable(A) to the category of $L_{\mathscr{C}}^f$-local spectra (i.e., spectra of the form $L_{\mathscr{C}}^f X$), it is left adjoint to the inclusion functor.

(b) There is a natural transformation $1 \to L_{\mathscr{C}}^f$.
(c) $L_{\mathscr{C}}^f$ is idempotent—for any X, the map $L_{\mathscr{C}}^f X \to L_{\mathscr{C}}^f L_{\mathscr{C}}^f X$ induced by the natural transformation in (b) is an equivalence.
(d) $L_{\mathscr{C}}^f$ is Bousfield localization with respect to the spectrum $L_{\mathscr{C}}^f S^0$.
(e) For any X, $L_{\mathscr{C}}^f X = X \wedge L_{\mathscr{C}}^f S^0$.
(f) For any finite X, $L_{\mathscr{C}}^f X = 0$ if and only if $X \in \mathscr{C}$. For any X, $L_{\mathscr{C}}^f X = 0$ if and only if X is in the localizing subcategory generated by $\mathscr{C}$.

Properties (a)–(c) say that $L_{\mathscr{C}}^f$ is a *localization* functor [**HPS97**, Definition 3.1.1]; property (e) says that $L_{\mathscr{C}}^f$ is *smashing* [**HPS97**, Definition 3.3.2]. We write $C_{\mathscr{C}}^f$ for the corresponding *acyclization* functor; that is, $C_{\mathscr{C}}^f X$ is the fiber of $X \to L_{\mathscr{C}}^f X$. Since $L_{\mathscr{C}}^f$ is smashing, then $C_{\mathscr{C}}^f X = X \wedge C_{\mathscr{C}}^f S^0$.

DEFINITION 3.1.2. For any slope n, let L_n^f denote finite localization away from the thick subcategory of finite spectra which are $\bigvee_{d \leq n} Z(d)$-acyclic, and write C_n^f for the corresponding acyclization.

In Section 2.5, we constructed a finite spectrum $F(u_1^{i_1}, \ldots, u_n^{i_n})$ for an appropriately chosen set of exponents $i_1, \ldots, i_n$, indexed by slopes. We will also write $F(U_n)$ for this spectrum. If d is the next largest slope after n, then by construction, $F(U_n)$ has a u_d-map. Let $\mathrm{Tel}(d)$ denote its telescope. Theorem 2.5.3 tells us that the Bousfield class $\langle F(U_n) \rangle$ is independent of the exponents $i_1, \ldots, i_n$; as noted after that theorem, the same is true of $\langle \mathrm{Tel}(d) \rangle$.

The following is based on Mahowald and Sadofsky's analysis in [**MS95**] of the functor L_n^f in the ordinary stable homotopy category.

PROPOSITION 3.1.3. *The functor L_n^f can be described in any of the following ways*:
(a) *as finite localization away from the thick subcategory generated by $F(U_n)$;*
(b) *as Bousfield localization with respect to $\bigvee_{d \leq n} \mathrm{Tel}(d)$;*
(c) *and as a colimit: if we let $\overline{F}(u_1^{i_1}, \ldots, u_n^{i_n})$ be the cofiber of projection to the top cell $F(u_1^{i_1}, \ldots, u_n^{i_n}) \to S^0$, then the map $X \to L_n^f X$ is given by*

$$X \to \varinjlim_{i_1, \ldots, i_n} X \wedge \overline{F}(u_1^{i_1}, \ldots, u_n^{i_n}).$$

(*The maps in the direct system will be defined in the proof.*)

Note also that for n large, L_n^f is not Bousfield localization with respect to $\bigvee_{d \leq n} Z(d)$. See Proposition 3.1.10.

PROOF. By Theorem 2.5.3(a), the thick subcategory of finite $\bigvee_{d \leq n} Z(d)$-acyclic spectra is equal to the thick subcategory generated by $F(u_1^{i_1}, \ldots, u_n^{i_n})$ for any choice of exponents $i_1, \ldots, i_n$. This proves part (a).

Part (b) is essentially a Bousfield class computation. Repeated application of Proposition 1.5.1(a) gives us

(3.1.4) $\qquad \langle S^0 \rangle = \langle F(U_n) \rangle \vee \langle \mathrm{Tel}(1) \rangle \vee \cdots \vee \langle \mathrm{Tel}(n) \rangle,$

as well as pairwise orthogonality of the Bousfield classes on the right. We need to show that a spectrum X is $\bigvee_{d \leq n} \mathrm{Tel}(d)$-acyclic if and only if it is L_n^f-acyclic. Assume that $\bigvee_{d \leq n} \mathrm{Tel}(d)_{**} X = 0$. By the decomposition of Bousfield classes (3.1.4), we see that $\langle X \rangle = \langle X \wedge F(U_n) \rangle$, so $L_n^f X = 0$ if and only if $L_n^f X \wedge F(U_n) = 0$. But certainly $L_n^f F(U_n) = 0$, so since L_n^f is smashing, then $L_n^f F(U_n) \wedge X = 0$. To show the converse, it suffices to show that $F(U_n)$ is $\bigvee_{d \le n} \mathrm{Tel}(d)$-acyclic; this follows by the orthogonality of the above Bousfield classes.

To prove (c), we first need to construct the direct system. By "projection to the top cell," we mean the composite $F(U_n) \to \cdots \to F(U_1) \to F(U_0) = S^0$. Using properties of the spectra $F(U_m)$ and their u_j-maps (Theorem 2.5.3), we can choose the u_j-maps compatibly; i.e., if we have exponents i_j and ℓ_j so that the objects $F(u_1^{i_1}, \ldots, u_n^{i_n})$ and $F(u_1^{\ell_1}, \ldots, u_n^{\ell_n})$ are defined and with $\ell_j \gg i_j$, then there is a map
$$F(u_1^{i_1}, \ldots, u_n^{i_n}) \to F(u_1^{\ell_1}, \ldots, u_n^{\ell_n})$$
which commutes with projection to the top cell. We let $\overline{F}(u_1^{i_1}, \ldots, u_n^{i_n})$ denote the cofiber of $F(u_1^{i_1}, \ldots, u_n^{i_n}) \to S^0$, and we define the spectrum $L_n' X$ to be the following colimit:

$$\begin{array}{ccc} X & \longrightarrow & \varinjlim_{i_1,\ldots,i_n} \overline{F}(u_1^{i_1}, \ldots, u_n^{i_n}) \wedge X \\ \parallel & & \parallel \\ X & \xrightarrow{g'} & L_n' X. \end{array}$$

For example, if $n = 1$, then we have $S^0 \xrightarrow{g'} u_1^{-1} S^0$, where $u_1^{-1} S^0$ is the mapping telescope of $u_1 \colon S^0 \to S^0$, and $\mathrm{fiber}(g') = F(u_1^\infty)$ is the analogue of the mod p^∞ Moore spectrum.

We need to verify that L_n' agrees with L_n^f. To do this, we note that L_n' is smashing, we show that $g' \colon S^0 \to L_n' S^0$ is an $L_n^f S^0$-equivalence, and we show that $L_n' S^0$ is $L_n^f S^0$-local. By construction, the fiber of g' is in the localizing subcategory generated by the spectra $F(u_1^{i_1}, \ldots, u_n^{i_n})$, and hence g' is an $L_n^f S^0$-equivalence.

To show that $L_n' S^0$ is local, we have to show that $[W, L_n' S^0]_{**} = 0$ for any L_n^f-acyclic W. For any finite localization, the acyclics are the localizing subcategory generated by the finite acyclics, so it suffices to show this when W is a finite acyclic. Note that if W is a finite acyclic, then so is DW, the Spanier-Whitehead dual of W. Since
$$[W, L_n' S^0]_{**} = [S^0, DW \wedge L_n' S^0]_{**} = \pi_{**} L_n' DW,$$
it suffices to show that $L_n' W = 0$ for any finite acyclic W, so let W be an acyclic. For each slope $j \le n$, if k is the next largest slope after j, then by vanishing lines, every u_k-map of $F(u_1^{i_1}, \ldots, u_j^{i_j}) \wedge W$ is nilpotent. Thus for all large enough exponents, $F(u_1^{i_1}, \ldots, u_n^{i_n}) \wedge W$ is equivalent to a wedge of suspensions of W. One can then argue as in [**MS95**, Proposition 3.8] to conclude that
$$\varinjlim_{i_1,\ldots,i_n} F(u_1^{i_1}, \ldots, u_n^{i_n}) \wedge W \to W$$
is an equivalence; hence the cofiber $L_n' W$ of this map is zero. $\square$

Recall from Definition 2.4.2 that a spectrum is of type n if it is $\bigvee_{d<n} Z(d)_{**}$-acyclic, but not $Z(n)_{**}$-acyclic. By Theorem 2.4.3, any finite type n spectrum has a u_n-map.

PROPOSITION 3.1.5. *Suppose that X is a finite type n spectrum with u_n-map u. Then $L_n^f X = u^{-1} X$.*

PROOF. As in the proof of Proposition 3.1.3(c), we have to check that the map $X \to u^{-1}X$ is an L_n^f-equivalence, and that $u^{-1}X$ is L_n^f-local. For the former, note that we have a diagram of cofibrations

$$\begin{array}{ccccc}
X/(u) & \longrightarrow & X & \xrightarrow{u} & X \\
\downarrow & & \| & & \downarrow u \\
X/(u^2) & \longrightarrow & X & \xrightarrow{u^2} & X \\
\downarrow & & \| & & \downarrow u \\
X/(u^3) & \longrightarrow & X & \xrightarrow{u^3} & X \\
\downarrow & & \| & & \downarrow u \\
\vdots & & \vdots & & \vdots
\end{array}$$

The sequential colimit is the cofibration

$$X/(u^\infty) \to X \to u^{-1}X.$$

Note that each spectrum $X/(u^n)$ is finite of type greater than n, and so is in the thick subcategory generated by $F(U_n)$ by Theorem 2.5.3(a). Hence $X/(u^\infty)$ is in $\text{loc}(F(U_n))$; hence $X \to u^{-1}X$ is an L_n^f-equivalence.

To show that $u^{-1}X$ is L_n^f-local, we have to show that $[F(U_n), u^{-1}X]_{**} = 0$. We compute:

$$[F(U_n), u^{-1}X]_{**} = \pi_{**}(DF(U_n) \wedge u^{-1}X)$$
$$= \pi_{**}(u^{-1}DF(U_n) \wedge X)$$
$$= u^{-1}\pi_{**}(DF(U_n) \wedge X).$$

Since $F(U_n)$ is of type greater than n, it has a vanishing line of slope greater than n; the same then goes for $DF(U_n) \wedge X$. Since u acts at slope n, it acts nilpotently on $\pi_{**}(DF(U_n) \wedge X)$. □

Here is the main theorem of this section. It tells us that L_n^f "kills off" certain P_t^s-homology groups; we also have a connectivity result.

THEOREM 3.1.6 (Margolis' killing construction). *Fix a slope n and let X be a spectrum.*

(a) *In the cofiber sequence $C_n^f X \xrightarrow{f} X \xrightarrow{g} L_n^f X$,*
 (i) $Z(d)_{**}g$ *is an isomorphism if $d \leq n$, and*
 (ii) $Z(d)_{**}f$ *is an isomorphism if $d > n$.*
In other words,
 (i)' $Z(d)_{**}C_n^f X = 0$ *if $d \leq n$, and*
 (ii)' $Z(d)_{**}L_n^f X = 0$ *if $d > n$.*
(b) *Suppose that X is a CL-spectrum (Definition 1.3.6). Then the homotopy of $L_n^f X$ is "bounded to the left": for each i, $\pi_{ij}L_n^f = 0$ for $j \ll 0$. If, in addition, X has finite type homotopy, then so does $L_n^f X$.*

PROOF. Part (a): By the theory of finite localization in [**HPS97**], the spectrum $C_n^f X$ is in the localizing subcategory generated by the finite $\bigvee_{d \leq n} Z(d)$-acyclics, and hence is $\bigvee_{d \leq n} Z(d)$-acyclic itself. This proves (i).

Since homology commutes with direct limits, we see that $Z(d)_{**}\operatorname{Tel}(k)$ is zero if $d \neq k$; in particular, $\bigvee_{k \leq n} \operatorname{Tel}(k)$ is $\bigvee_{d>n} Z(d)$-acyclic. Since L_n^f is Bousfield localization with respect to $\bigvee_{k \leq n} \operatorname{Tel}(k)$, then we conclude that $L_n^f S^0$ is $\bigvee_{d>n} Z(d)$-acyclic. Since L_n^f is smashing, this proves (ii).

Part (b): Given X as in the statement, we show that the maps in the direct system defining $L_n^f X$ (Proposition 3.1.3(c)) are isomorphisms on π_{**} in a range increasing with the exponents.

We let $F(U_k) = F(u_1^{i_1}, \ldots, u_k^{i_k})$ and we assume that d is the next largest slope after k. We write $F(U_k)(u_d^{i_d}) = F(u_1^{i_1}, \ldots, u_d^{i_d})$ for the fiber of the u_d-map on $F(U_k)$. Consider the following commutative diagram:

$$\begin{array}{ccccccc}
\cdots \longrightarrow & \Sigma^{-i_d+1,-i_d d}F(U_k) & \longrightarrow & F(U_k)(u_d^{i_d}) & \longrightarrow & F(U_k) & \xrightarrow{u_d^{i_d}} \cdots \\
& \downarrow u_d^{i_d} & & \downarrow f_1 & & \| & \\
\cdots \longrightarrow & \Sigma^{-2i_d+1,-2i_d d}F(U_k) & \longrightarrow & F(U_k)(u_d^{2i_d}) & \longrightarrow & F(U_k) & \xrightarrow{u_d^{2i_d}} \cdots \\
& \downarrow u_d^{i_d} & & \downarrow f_2 & & \| & \\
\cdots \longrightarrow & \Sigma^{-3i_d+1,-3i_d d}F(U_k) & \longrightarrow & F(U_k)(u_d^{3i_d}) & \longrightarrow & F(U_k) & \xrightarrow{u_d^{3i_d}} \cdots \\
& \downarrow u_d^{i_d} & & \downarrow f_3 & & \| & \\
& \vdots & & \vdots & & \vdots &
\end{array}$$

It suffices to show that the maps $f_r \wedge 1_X$ are isomorphisms in a range increasing with r. Well, $f_r \wedge 1_X$ is an isomorphism whenever the map

$$(3.1.7) \qquad \Sigma^{-ri_d+1,-ri_d d}F(U_k) \wedge X \xrightarrow{u_d^{i_d} \wedge 1_X} \Sigma^{-(r+1)i_d+1,-(r+1)i_d d}F(U_k) \wedge X$$

from the left column is. The fiber is $\Sigma^{-ri_d+1,-ri_d d}F(U_k)(u_d^{i_d}) \wedge X$; this spectrum has a vanishing line with some slope $\ell > d$, and we can use Remark 2.3.5 to compute its intercept. To do this, we compute $(H\mathbf{F}_p)_{**}F(U_k)(u_d^{i_d})$—each u_j-map induces the zero map on $(H\mathbf{F}_p)_{**}$, so this computation is easy. We find that

  (i) $(H\mathbf{F}_p)_{i*}F(U_k)(u_d^{i_d}) = 0$ for $i \ll 0$,
  (ii) $(H\mathbf{F}_p)_{i*}F(U_k)(u_d^{i_d}) = 0$ for $i > 0$, and
  (iii) $(H\mathbf{F}_p)_{*j}F(U_k)(u_d^{i_d}) = 0$ for $j < N$ for some number N.

By the Hurewicz Theorem 1.3.5, (i) and (iii) also hold for π_{**}; hence we can compute the equation of the vanishing line for $F(U_k)(u_d^{i_d})$. Smashing with X moves the intercept a bit, so we find that $\pi_{ij}F(U_k)(u_d^{i_d}) \wedge X = 0$ when $j < \ell i + N'$ for some N'. Therefore the map (3.1.7) is an isomorphism on π_{ij} when $j + ri_d d < \ell(i + ri_d - 1) + N'$, i.e., when $j < \ell i + ri_d(\ell - d) + (N' - \ell)$. Since $\ell > d$, then as r increases, this range increases.

So we see that for each s, then the graded groups π_{s*} of the direct system have a uniform bound to the left; hence the same is true of the direct limit. If X also has finite type homotopy, then since the homotopy of each $F(U_k)$ is of finite type, the same goes for $F(U_k) \wedge X$; since the homotopy of the direct system stabilizes at each bidegree, we are done. $\square$

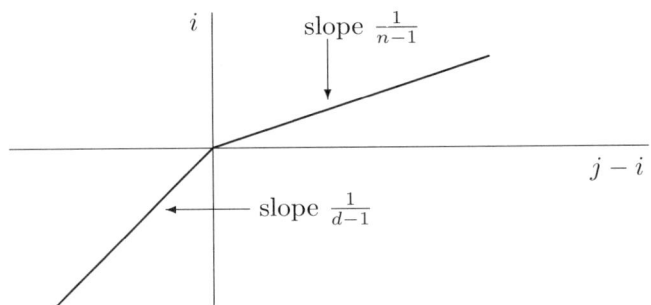

FIGURE 3.1.A. "Vanishing curve" for $\pi_{ij}(C_n^f S^0)$: $\pi_{ij}(C_n^f S^0)$ is zero above the two indicated lines. Here, d is the slope preceding n, and slopes are labeled in $(i, j-i)$-coordinates. These lines are drawn through the origin for convenience, but their intercepts may be nonzero.

REMARK 3.1.8. (a) Much of this was already known in the module setting. For $p = 2$, this is due to Margolis [**Mar83**, Theorem 21.1]; for arbitrary primes, part (a) and the connectivity in part (b) can be found in [**Pal92**, Theorem 3.1]. Given an A-comodule M and a slope n, then the module-version of the killing construction (dualized to the category of comodules) gives comodules $M\langle n+1, \infty\rangle$ and $M\langle 1, n\rangle$, well-defined up to injective summands, and an injective comodule J so that the following is short exact:
$$0 \to M\langle n+1, \infty\rangle \to M \oplus J \to M\langle 1, n\rangle \to 0.$$
Being well-defined up to injective summands translates in our setting to being well-defined up to an object of $\text{loc}(A)$ (cf. Lemma 3.1.9). So if X is an injective resolution of M, then an injective resolution for $M\langle 1, n\rangle$ is a connective cover of $L_n^f X$.

(b) By studying the proof of Theorem 3.1.6(b), one can show that $C_n^f S^0$ has a nice vanishing curve: for $i \geq 0$, $\pi_{ij} C_n^f S^0$ has a vanishing line of slope n, and if d is the slope preceding n, then for $i < 0$, $\pi_{ij} C_n^f S^0$ has a vanishing line of slope d. For example, when $n = 1$, we have

$$\begin{array}{ccccc} C_1^f S^0 & \longrightarrow & S^0 & \longrightarrow & L_1^f S^0 \\ \| & & \| & & \| \\ F(u_1^\infty) & \longrightarrow & S^0 & \longrightarrow & u_1^{-1} S^0. \end{array}$$

So
$$\pi_{**} L_1^f S^0 = u_1^{-1} \pi_{**} S^0 = h_{10}^{-1} \pi_{**} S^0 = \mathbf{F}_2[h_{10}^{\pm 1}].$$

Hence $\pi_{ij} C_1^f S^0$ has a vanishing line of slope 2 for positive i, and a vanishing line of slope 1 (a vertical line, in $(i, j-i)$-coordinates) for i negative. See also Figure 3.1.A.

We need the following property of L_n^f in the next section.

LEMMA 3.1.9. *For any slope n, $L_n^f A = 0$. Hence any spectrum X in the localizing subcategory generated by A is L_n^f-acyclic.*

PROOF. Since the category of L_n^f-acyclics contains a finite spectrum $F(U_n)$, then it contains A by Lemma 1.3.9. □

We close this section with a few remarks about the telescope conjecture (see [**Rav84**] and [**HPS97**, Definition 3.3.8]). Let L_n denote Bousfield localization with respect to $\bigvee_{d \leq n} Z(d)$.

PROPOSITION 3.1.10. *Let $p = 2$. We have $L_1 = L_1^f$. If $n > 1$, then $L_n \neq L_n^f$.*

One can interpret the statement "$L_n = L_n^f$" as an analogue of the telescope conjecture, so this result would say that the telescope conjecture fails except when $n = 1$. However, it seems more proper to refer to the statement "every smashing localization functor is a finite localization" as the telescope conjecture—see [**HPS97**, Definition 3.3.8]. Since we do not know whether L_n is smashing, this result does not necessarily present a counterexample to this version of the telescope conjecture.

PROOF. We know that the sphere has a u_1-map; a simple computation shows that $u_1^{-1} S^0 = Z(1)$, so that Bousfield localization with respect to $L_1^f S^0 = u_1^{-1} S^0$ is the same as that with respect to $Z(1)$.

There is a non-nilpotent element $d_0 \in \pi_{4,18} S^0 = \operatorname{Ext}_A^{4,18}(\mathbf{F}_2, \mathbf{F}_2)$. We claim that $d_0^{-1} S^0$ is L_n-acyclic for all n, and L_n^f-local for $n \geq 3$. Since 3 is the next slope after 1, this covers all of the possibilities.

For degree reasons, d_0 induces the zero map on $Z(d)_{**}$ for all d. (If an element $\alpha \in \pi_{ij} S^0$ is nonzero on $Z(d)_{**} S^0 = Z(d)_{**}$, then j/i must be a multiple of d.) Hence $Z(d)_{**}(d_0^{-1} S^0) = 0$ for all d, and $d_0^{-1} S^0$ is L_n-acyclic for all n. To show that $d_0^{-1} S^0$ is L_n^f-local, we show that $[F, d_0^{-1} S^0]_{**} = 0$ for any finite $\bigvee_{d \leq n} Z(d)$-acyclic F. We compute:

$$[F, d_0^{-1} S^0]_{**} = [S^0, DF \wedge d_0^{-1} S^0]_{**}$$
$$= [S^0, d_0^{-1} DF]_{**}$$
$$= d_0^{-1} \pi_{**} DF.$$

Since F is a finite $\bigvee_{d \leq n} Z(d)_{**}$-acyclic, so is DF (by Proposition 2.2.10). By Theorem 2.3.1, $\pi_{**} DF$ has a vanishing line of some slope $m > n$. Since $n \geq 3$, we know that $m \geq 6$, so the slope m is larger than $\frac{9}{2}$, the slope of d_0. Hence d_0 acts nilpotently on $\pi_{**} DF$; i.e., $d_0^{-1} \pi_{**} DF = 0$. □

When p is odd, one can again show that $L_1 = L_1^f$. There is every reason to expect that there are non-nilpotent elements in $\operatorname{Ext}_A^{**}(\mathbf{F}_p, \mathbf{F}_p)$ which are not detected by any single $Z(d)$; given this, one could conclude that $L_n \neq L_n^f$ for n large.

3.2. A Tate version of the functor H

In this section we introduce a "Tate" version of the functor H, defined for finite-dimensional quotients of A, and we show that when n is large enough, then $L_n^f HB = \widehat{H}B$. We will use our computations here to prove chromatic convergence in Section 3.3.

Let B be a quotient Hopf algebra of A. Note that practically all of our results hold in the category $\mathsf{Stable}(B)$; the only exceptions are computational results like the strict inequalities in Theorem 6.6.1. Let $\operatorname{res}_{A,B} \colon \mathsf{Stable}(A) \to \mathsf{Stable}(B)$ denote the forgetful functor (also known as *restriction*); $\operatorname{res}_{A,B}$ has a right adjoint, *induction*, written $\operatorname{ind}_{B,A}$, and defined by $X \mapsto A \square_B X$ (cf. Section 1.2—in this section, we use the notations $\operatorname{res}_{A,B}$ and $\operatorname{ind}_{B,A}$ rather than $\downarrow_B$ and $A\,\square_B-$). These functors are exact (i.e., they take cofibrations to cofibrations), and restriction preserves the smash product and the sphere object. In the language of [**HPS97**, Section 3.4], the restriction functor is a *stable morphism*.

When B is a finite-dimensional Hopf algebra, we may consider another stable homotopy category associated to it, the *stable category of B-comodules*, written $\mathsf{StComod}(B)$. We provide a brief review of the relevant results here; see [**HPS97**, Section 9.6] for a few more details. The objects of $\mathsf{StComod}(B)$ are B-comodules, and the morphisms $\underline{\operatorname{Hom}}_B^*$ are defined as follows: define the morphisms of degree zero to be $\underline{\operatorname{Hom}}_B(X,Y) = \operatorname{Hom}_B(X,Y)/\simeq$, where $f \simeq g\colon X \to Y$ if $f - g$ factors through an injective comodule. Hence two comodules are equivalent in $\mathsf{StComod}(B)$ if they differ by injective summands. Define the desuspension functor $\Sigma^{-1} = \Sigma^{-1,0}$ by the short exact sequence

$$0 \to X \xrightarrow{\eta \wedge 1_X} B \otimes X \to \Sigma^{-1} X \to 0,$$

where $B \otimes X$ is the cofree comodule on X (Definition 0.4.1). This functor is invertible: ΣX is any comodule which fits into a short exact sequence

$$0 \to \Sigma X \to I \to X \to 0,$$

where I is injective. (Since B is finite, then injectives and projectives are the same, and there are enough projectives; hence there is always a surjection $I \twoheadrightarrow X$.) We let

$$\underline{\operatorname{Hom}}_B^i(X,Y) = \underline{\operatorname{Hom}}_B(\Sigma^i X, Y).$$

As with the category $\mathsf{Stable}(B)$, this category is bigraded; thus the suspension functor has a second, "internal," index, as do sets of morphisms.

$\mathsf{StComod}(B)$ is a stable homotopy category. Indeed, on $\mathsf{Stable}(B)$ one has finite localization (Definition 3.1.1) away from the thick subcategory generated by the finite spectrum $B = H\mathbf{F}_p$; $\mathsf{StComod}(B)$ is equivalent to the full subcategory of $\mathsf{Stable}(B)$ of L_B^f-local objects—see [**HPS97**, Theorems 9.6.3–9.6.4]. So we have a functor $L_B^f\colon \mathsf{Stable}(B) \to \mathsf{StComod}(B)$ with a right adjoint $J\colon \mathsf{StComod}(B) \to \mathsf{Stable}(B)$. (Every localization functor has a right adjoint, namely inclusion of the local objects into the category.)

DEFINITION 3.2.1. Let B be a quotient Hopf algebra of A. We note that the object HB in $\mathsf{Stable}(A)$ is $\operatorname{ind}_{B,A}(S^0)$, and when B is finite-dimensional, we define the object $\widehat{HB}$ in $\mathsf{Stable}(A)$ to be $\operatorname{ind}_{B,A}(J(S^0))$.

Fix a slope n. On each of the categories $\mathsf{Stable}(A)$, $\mathsf{Stable}(B)$, and $\mathsf{StComod}(B)$, we can define the functor L_n^f to be smashing with $L_n^f S^0$, or with its image under res or $L_B^f \circ \operatorname{res}$. This gives us a commuting diagram of functors

$$\begin{array}{ccccc}
\mathsf{Stable}(A) & \xrightarrow{\operatorname{res}} & \mathsf{Stable}(B) & \xrightarrow{L_B^f} & \mathsf{StComod}(B) \\
\downarrow{L_n^f} & & \downarrow{L_n^f} & & \downarrow{L_n^f} \\
L_n^f\mathsf{Stable}(A) & \xrightarrow{\operatorname{res}} & L_n^f\mathsf{Stable}(B) & \xrightarrow{L_B^f} & L_n^f\mathsf{StComod}(B),
\end{array}$$

along with the commuting diagram of their right adjoints

$$\begin{array}{ccccc}
\mathsf{Stable}(A) & \xleftarrow{\mathrm{ind}} & \mathsf{Stable}(B) & \xleftarrow{J} & \mathsf{StComod}(B) \\
\uparrow R_{\mathsf{Stable}(A)} & & \uparrow R_{\mathsf{Stable}(B)} & & \uparrow R_{\mathsf{St}(B)} \\
L_n^f \mathsf{Stable}(A) & \xleftarrow{\mathrm{ind}} & L_n^f \mathsf{Stable}(B) & \xleftarrow{J} & L_n^f \mathsf{StComod}(B).
\end{array}$$

(The first diagram commutes by the definition of the vertical functors, and the second diagram commutes as a result—given a commuting square of left adjoints, their right adjoints also commute.) Recall from Notation 2.2.8 that y_d is the element of A—either $\xi_t^{p^s}$ or τ_n—"with slope d."

PROPOSITION 3.2.2. *Fix a finite-dimensional quotient Hopf algebra B of A, and let*

$$N = \max\{d \mid y_d \neq 0 \text{ in } B\}.$$

If n is a slope larger than N, then $L_n^f HB = \widehat{H}B$.

PROOF. We claim that the two maps

$$\alpha \colon HB \to L_n^f HB,$$
$$\beta \colon HB \to \widehat{H}B,$$

are the same. To show this, we show that the spectrum $\widehat{H}B$ is L_n^f-local, and that the map $\beta \colon HB \to \widehat{H}B$ is an L_n^f-equivalence. By Lemma 3.2.4 below, the fiber of β has homotopy concentrated in the third quadrant, so by Lemma 1.3.8, it is in $\mathrm{loc}(A)$. In particular, it is L_n^f-acyclic by Lemma 3.1.9; hence β is an L_n^f-equivalence.

To show that $\widehat{H}B$ is local, we show that it is in the image of the right adjoint $R_{\mathsf{Stable}(A)}$ of $L_n^f \colon \mathsf{Stable}(A) \to L_n^f \mathsf{Stable}(A)$. We claim, in fact, that $R_{\mathsf{St}(B)}$ is the identity functor. Given this claim, then we use the commutativity of the diagram of right adjoints: we have

$$\widehat{H}B = \mathrm{ind}(JS^0)$$
$$= \mathrm{ind}(J(R_{\mathsf{St}(B)}S^0))$$
$$= R_{\mathsf{Stable}(A)}(\mathrm{ind}(JS^0)).$$

Hence $\widehat{H}B$ is in the image of $R_{\mathsf{Stable}(A)}$, and so it is local.

It remains to verify the claim that $R_{\mathsf{St}(B)}$ is the identity functor, or what is the same, that

$$L_n^f \colon \mathsf{StComod}(B) \to \mathsf{StComod}(B)$$

is the identity functor. So it suffices to show that the cofiber of $S^0 \to L_n^f S^0$ is zero in the category $\mathsf{StComod}(B)$. This cofibration is obtained by applying the functor $L_B^f \circ \mathrm{res}$ to the following cofibration in $\mathsf{Stable}(A)$:

$$C_n^f S^0 \to S^0 \to L_n^f S^0.$$

So the cofiber under consideration is $L_B^f(\mathrm{res}\, C_n^f S^0)$. By Proposition 3.1.3, L_n^f is finite-localization away from the thick subcategory generated by $F(U_n)$; but $F(U_n)$ has no $Z(d)$-homology for any d with $y_d \in B$, so $\mathrm{res}(F(U_n))$ is in $\mathrm{loc}(B)$. Since $\mathrm{res}(F(U_n))$ is finite, it is in fact in $\mathrm{thick}(B)$. By fundamental properties of finite localizations (Definition 3.1.1), $\mathrm{res}(C_n^f S^0)$ is in $\mathrm{loc}(F(U_n))$, and hence in $\mathrm{loc}(B)$. Hence $L_B^f(\mathrm{res}\, C_n^f S^0) = 0$, as desired. □

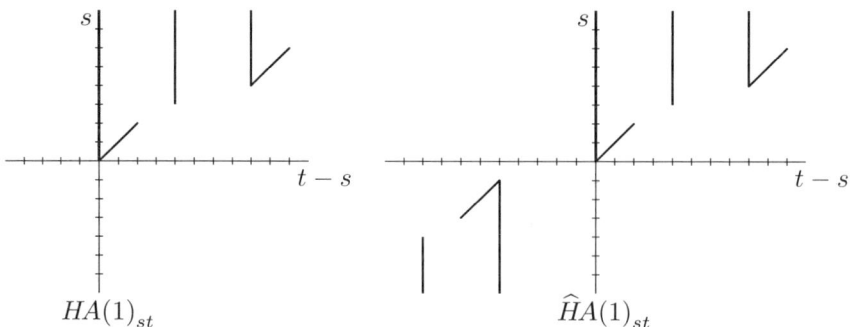

FIGURE 3.2.A. The coefficients of $HA(1)$ and $\widehat{H}A(1)$ when $p = 2$. Since the top cell of $A(1)$ is in degree 6, the third quadrant of $\widehat{H}A(1)_{**}$ is the same as the first quadrant, reflected across the origin and then translated by $(-1, -6)$.

We are particularly interested in the case $B = A(m)$, the quotient of A defined in Example 2.1.4.

COROLLARY 3.2.3. *Fix an integer $m \geq 0$. For n sufficiently large, we have*
$$L_n^f HA(m) = \widehat{H}A(m).$$

In particular, n should be larger than
$$\max\{d \mid y_d \neq 0 \text{ in } A(m)\} = \begin{cases} |\xi_{m+1}| = 2^{m+1} - 1, & \text{if } p = 2, \\ \frac{p}{2}|\xi_m| = p^{m+1} - p, & \text{if } p \text{ is odd}. \end{cases}$$

Now we "compute" the homotopy groups of $\widehat{H}B$. See Figure 3.2.A for an example.

LEMMA 3.2.4. *Let B be a finite-dimensional quotient Hopf algebra of A, and consider $\pi_{ij}\widehat{H}B$.*
  (a) *When $ij < 0$, then $\pi_{ij}\widehat{H}B = 0$. (In other words, the homotopy is concentrated in the first and third quadrants.)*
  (b) *When i and j are nonnegative, the map $HB \to \widehat{H}B$ induces an isomorphism $\pi_{ij}\widehat{H}B \cong \pi_{ij}HB$.*
  (c) *Let d be the maximal degree in which B is nonzero. When i and j are negative, we have an isomorphism $\pi_{ij}\widehat{H}B \cong \pi_{-1-i,-d-j}HB$.*

PROOF. We work in the category $\mathsf{Stable}(B)$. We have adjoint functors
$$L\colon \mathsf{Stable}(B) \to \mathsf{StComod}(B),$$
$$J\colon \mathsf{StComod}(B) \to \mathsf{Stable}(B),$$
and we want to compute the homotopy of JS^0 by comparing it to that of $S^0 \in \mathsf{Stable}(B)$. To do this, we need to recall the description of the functor J from [**HPS97**, Lemma 9.6.7]. Let I denote an injective resolution of the B-comodule $\mathbf{F}_p$ (i.e., $I \cong S^0$ in $\mathsf{Stable}(B)$); then $\mathrm{Hom}_{\mathbf{F}_p}(I, \mathbf{F}_p)$ is a projective resolution of $\mathbf{F}_p$. We splice these resolutions together to get the "Tate complex" $t_B(\mathbf{F}_p)$:
$$\cdots \to \mathrm{Hom}(I_1, \mathbf{F}_p) \to \mathrm{Hom}(I_0, \mathbf{F}_p) \to I_0 \to I_1 \to \cdots.$$

Then the functor J is defined by $J(M) = t_B(\mathbf{F}_p) \otimes M$. It is clear that the map $S^0 \to J(S^0)$ induces an isomorphism on π_{i*} for $i \geq 0$. If we take I to be a minimal injective resolution, then $\mathrm{Hom}(I, \mathbf{F}_p)$ is a minimal projective resolution; by minimality, $[S^0, JS^0]_{i*} = [\mathbf{F}_p, JS^0]_{i*}$ gives the primitives in JS^0 in degree i, and there is one primitive for each summand isomorphic to B. Since $\mathrm{Hom}(B, \mathbf{F}_p) \cong \Sigma^{-d} B$, we get the reflection and translation as described in (c). □

3.3. Chromatic convergence

It is easy to see that given a spectrum X and slopes n_1 and n_2 with $n_1 < n_2$, then the map $X \to L_{n_1}^f X$ factors through $X \to L_{n_2}^f X$. Hence we get a tower of cofibrations:

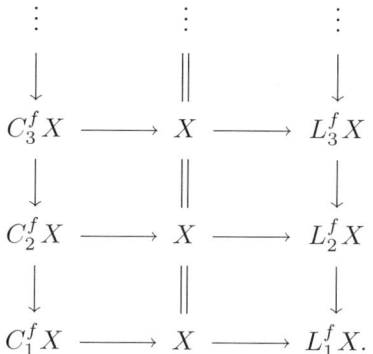

(Strictly speaking, we have defined L_n^f only when n is a slope. For a general n, we define L_n^f, as in Definition 3.1.2, to be finite localization away from the finite spectra which are $\bigvee_{d \leq n} Z(d)$-acyclic. So if n is not a slope, then $L_n^f = L_{n-1}^f$.) We may as well just focus on the right hand column, giving the following diagram:

$$0 \longleftarrow L_1^f X \longleftarrow L_2^f X \longleftarrow L_3^f X \longleftarrow \cdots.$$
$$=\uparrow \quad\quad \uparrow \quad\quad \uparrow$$
$$M_1^f X \quad\quad M_2^f X \quad\quad M_3^f X$$

Here, $M_n^f X = X \wedge M_n^f S^0$ is defined to be the fiber of $L_n^f X \to L_{n-1}^f X$. We call this diagram the *chromatic tower* for X. (Note that if n is not a slope, then $M_n^f X = 0$.)

Margolis [**Mar83**, Theorem 22.4] proved the following theorem for A^*-modules when $p = 2$; see also [**Pal94**].

THEOREM 3.3.1 (Chromatic convergence). *If X is finite, then $X = \varprojlim_n L_n^f X$.*

Indeed, the proof shows that the tower of groups $\pi_{**} L_n^f X$ is pro-constant. In order to prove Theorem 3.3.1, we need the following observation.

COROLLARY 3.3.2. *For any m, the chromatic tower for $HA(m)$ stabilizes to $\widehat{HA}(m)$.*

PROOF. This follows from Corollary 3.2.3. Actually, that result merely says that $L_n^f HA(m) = \widehat{HA}(m)$ for all sufficiently large n; it does not say that the map $L_{n+1}^f HA(m) \to L_n^f HA(m)$ is an equivalence when $n \gg 0$. This follows from the proof of Proposition 3.2.2, though. □

Note that the chromatic tower for $HA(m)$ converges to $\widehat{HA}(m)$, not to $HA(m)$—we do not have chromatic convergence in this case.

PROOF OF THEOREM 3.3.1. If X is finite, then smashing with X commutes with inverse limits. Since $L_n^f X = X \wedge L_n^f S^0$, then chromatic convergence for S^0 implies chromatic convergence for all finite spectra.

By Theorem 3.1.6(b), $L_n^f S^0$ has finite type homotopy bounded to the left, so by Proposition 2.1.5, we see that
$$L_n^f S^0 = \varprojlim_m HA(m) \wedge L_n^f S^0 = \varprojlim_m L_n^f HA(m).$$

Hence we have
$$\varprojlim_n L_n^f S^0 = \varprojlim_n (\varprojlim_m L_n^f HA(m))$$
$$= \varprojlim_m (\varprojlim_n L_n^f HA(m))$$
$$= \varprojlim_m \widehat{HA}(m).$$

Now we use Lemma 3.2.4 to compute the homotopy groups of $\varprojlim \widehat{HA}(m)$: in the first quadrant, the inverse limit stabilizes in each bidegree to the homotopy of $\varprojlim HA(m) = S^0$, by Proposition 2.1.5. In the second and fourth quadrants, there is no homotopy at any stage in the inverse system. In the third quadrant, for each bidegree (i, j) there is an m_0 so that $\pi_{ij}\widehat{HA}(m) = 0$ for all $m \geq m_0$; so the inverse limit has no homotopy in the third quadrant, either. Hence the map $S^0 \to \varprojlim \widehat{HA}(m)$ is an isomorphism in homotopy. $\square$

One could probably get another proof of chromatic convergence by making precise, and then using, the vanishing curve result mentioned in Remark 3.1.8(b).

By the way, Margolis gets chromatic convergence in his setting for all bounded below modules; since we do not have convergence for $HA(n)$ (which is an injective resolution of a bounded below comodule), we cannot expect our result to generalize, precisely as stated, to other spectra. On the other hand, perhaps one could modify the proof of Theorem 3.3.1 to show that for any CL-spectrum X, then $X \to \varprojlim L_n^f X$ is a connective cover. For this, the alternate formulation of CL after Definition 1.3.6 might be useful—X is CL if and only if there are numbers i_0, i_1, and j_0 so that X has a cellular tower built of spheres $S^{i,j}$ with $i_0 \leq i \leq i_1$ and $j - i \geq j_0$.

3.4. Further discussion: work of Mahowald and Shick

Chromatic convergence for the sphere gives a convergent filtration of $\pi_{**}S^0 = \operatorname{Ext}_A^{**}(\mathbf{F}_p, \mathbf{F}_p)$. When $p = 2$, Mahowald and Shick [**MS87**] give another convergent filtration; are these filtrations the same? If $p = 2$, the slope of the element ξ_{n+1} is $2^{n+1} - 1$; this element corresponds to the periodicity operator $v_n = h_{n+1,0}$. Mahowald and Shick also construct something they call $v_n^{-1} \operatorname{Ext}_A^{**}(\mathbf{F}_2, \mathbf{F}_2)$. How does this compare to $\pi_{**}(L_{2^{n+1}-1}^f S^0)$? (Shick [**Shi88**] has done similar work at odd primes, and one can naturally ask the same questions about that.) We describe their work. When $p = 2$, we write v_n for $h_{n+1,0}$, and we write E for the quotient $\mathbf{F}_2[\xi_{n+1}]/(\xi_{n+1}^2)$; when p is odd, we write E for $E[\tau_n]$. They show that for every $k \geq n$, the restriction map
$$\operatorname{Ext}_{A(k)}^{**}(\mathbf{F}_p, \mathbf{F}_p) \to \operatorname{Ext}_E^{**}(\mathbf{F}_p, \mathbf{F}_p) \cong \mathbf{F}_p[v_n]$$

is nontrivial, so some power of v_n lifts to $\text{Ext}^{**}_{A(k)}(\mathbf{F}_p, \mathbf{F}_p)$ (this is a special case of Proposition 2.4.5); furthermore, one can choose the lifts compatibly as k varies. Hence the map
$$\text{Ext}^{**}_{A(k+1)}(\mathbf{F}_p, \mathbf{F}_p) \to \text{Ext}^{**}_{A(k)}(\mathbf{F}_p, \mathbf{F}_p)$$
induces a map after inverting v_n. They define
$$v_n^{-1} \text{Ext}^{**}_A(\mathbf{F}_p, \mathbf{F}_p) = \varprojlim_k (v_n^{-1} \text{Ext}^{**}_{A(k)}(\mathbf{F}_p, \mathbf{F}_p)).$$

From our point of view, the lift of v_n to $\text{Ext}^{**}_{A(k)}(\mathbf{F}_p, \mathbf{F}_p)$ is an element in $\pi_{**}HA(k)$. Given a homotopy class α in a ring spectrum R, one can turn it into a self-map by the composite
$$R \xrightarrow{\eta \wedge 1} S^0 \wedge R \xrightarrow{\alpha \wedge 1} R \wedge R \xrightarrow{\mu} R.$$
Therefore we can invert the self-map v_n of $HA(k)$; Mahowald and Shick's result says that we can do this compatibly, giving maps of spectra
$$v_n^{-1} HA(k+1) \to v_n^{-1} HA(k).$$
Since the inverse limit of the rings $HA(k+1)$ is the sphere, by Proposition 2.1.5, then it seems plausible to take the inverse limit of the objects $v_n^{-1} HA(k)$ to define $v_n^{-1} S^0$.

More generally, if X is any spectrum, one has (at least) two choices for the definition of $v_n^{-1} X$: either
$$X \wedge v_n^{-1} S^0 = X \wedge \varprojlim_k (v_n^{-1} HA(k)),$$
or
$$\varprojlim_k (X \wedge v_n^{-1} HA(k)).$$
If X is finite, then these are the same; we write $v_n^{-1} X$ to denote either of them. Otherwise, though, they need not be the same, and it is not clear which is the better choice. The first makes the functor $v_n^{-1}(-)$ smashing, which could be an advantage.

Let a_n be the slope of v_n, either $2^{n+1} - 1$ or $2p^n - 1$, depending on whether p is even or odd. When X is finite, one can see that $v_n^{-1} X$ is $L_{a_n}^f$-local, so the map $X \to v_n^{-1} X$ factors through $L_{a_n}^f X$, giving a map
$$L_{a_n}^f X \to v_n^{-1} X.$$
Is this map an equivalence?

3.5. Further discussion

It seems most convenient to consider the chromatic tower as above, in which we kill off the P_t^s-homology groups of X in order of slope. This ordering is what allows us to prove chromatic convergence, via the "connectivity result" of Theorem 3.1.6(b). But the results of Chapters 5 and 6 suggest another approach, at least at the prime 2. There is a quotient Hopf algebra D of A, and the corresponding ring HD_{**} seems to play a very important role in the category $\mathsf{Stable}(A)$; instead of viewing the chromatic construction as killing off P_t^s-homology groups, one can view it as killing off generators of the ring HD_{**} in a particular order. As we describe in Section 6.9, though, when $p = 2$ we can kill off these generators in many different orders, and indeed, we can choose different generators entirely. With a different choice of generators and/or a different ordering, can one still prove convergence? Does other structure reveal itself when one works with other generators of HD_{**} or with other orderings of the same generators?

As noted after the statement of Proposition 3.1.10, the validity of the strong form of the telescope conjecture is not known in $\mathsf{Stable}(A)$. If one could show that Bousfield localization with respect to $\bigvee_{d \leq n} Z(d)$ were smashing, that would disprove it; we would be interested in any progress along these lines.

It is conceivable that the finite-type result of Theorem 3.1.6(b) could have bearing on the convergence (in the ordinary stable homotopy category) of the tower

$$\cdots \to L_2^f X \to L_1^f X \to L_0^f X.$$

After all, the v_n-localized Adams spectral sequence in [**MS95**] has an E_2-term with some relation to what we call $\pi_{**} L_m^f S^0$ for a suitable m, so computations and other information about $\pi_{**} L_m^f S^0$ could have applications to the finite localization functor L_n^f in ordinary stable homotopy theory.

Work of Mahowald and others has led to calculations similar to that of $\widehat{HA}(m)$ in Lemma 3.2.4; see [**MR**], for example. This may provide more connections between our functors L_n^f and topology.

In parallel with the study of chromatic phenomena in the ordinary stable homotopy category, one should try to understand the filtration pieces in the chromatic tower, and "monochromatic" objects in general: these are objects X so that $M_n^f X = X$. Other than the trivial case of $M_1^f X = h_{10}^{-1} X$, we have no information about these objects. See [**Pal94**] for some information about chromatic structure in the module setting.

Lastly, it seems possible that the chromatic tower constructed here, as well as those with other orderings, are the Steenrod algebra analogues of the multiple complexes of Benson and Carlson [**BC87**]. Is this a good analogy? If so, does it give any new insight into $\mathsf{Stable}(A)$, $\mathsf{Stable}(kG)$, or $\mathsf{Stable}(\Gamma)$ for an arbitrary commutative Hopf algebra Γ? Evens and Siegel [**ES96**] have extended the multiple complex construction to modules over finite-dimensional cocommutative Hopf algebras, or equivalently to comodules over finite-dimensional commutative Hopf algebras, so their work is relevant here.

CHAPTER 4

Computing Ext with elements inverted

In this chapter we present a spectral sequence for computing $v_n^{-1}\operatorname{Ext}_A^{**}(M,N)$ for appropriate comodules M and N, and we give several applications. The spectral sequence is a localized Adams spectral sequence; we prove convergence using vanishing planes and a genericity argument.

In this chapter, we use the following notation: when $p = 2$, then $q_n = p_{n+1}^0$, $Q_n = P_{n+1}^0$, and $v_n = h_{n+1,0}$. When p is odd, then q_n, Q_n, and v_n are as usual—see Section 2.2.

We start the chapter in Section 4.1 with a discussion of the q_n-based Adams spectral sequence. Since q_n is connective, convergence for this spectral sequence is tractable. q_n is a nice ring spectrum, and its coefficient ring is polynomial on a class v_n. In Section 4.2 we try to invert some power of this class in the q_n-based spectral sequence converging to $\pi_{**}X$. Depending on X, this may yield a new spectral sequence. When it does, we give a condition on the unlocalized spectral sequence— the existence of a vanishing plane with specified "slope"—that ensures convergence of the localized one. By appealing to Theorem 1.4.4, we see that convergence is generic in X.

Next, in Section 4.3 we present a spectrum X for which we can show that the E_2-term has an appropriate vanishing plane. It turns out that X can be built out of the sphere by a series of self-maps, similarly to the spectra $F(u_{d_1}^{j_1},\ldots,u_{d_n}^{j_n})$ of Section 2.5. This allows us to show that all finite spectra which are $L_{a_n}^f$-acyclic satisfy the convergence condition for the localized spectral sequence, where a_n is the slope of v_n. We remark that X is the injective resolution of the comodule known as $\frac{1}{2}A(n)$, so our discussion of X involves an investigation of the comodule structure on $A(n)$ which was constructed by Mitchell [**Mit85**] and Smith [**Smi**]. This discussion spans Sections 4.3–4.5.

We close the chapter with Section 4.6, in which we do some computations and discuss some applications. First we compute $(Q_n)_{**}(Q_n)$ for all n, and from this $\operatorname{Ext}_{(Q_0)_{**}(Q_0)}^{**}((Q_0)_{**}, (Q_0)_{**}X)$ for every connective X. This lets us compute $v_0^{-1}\pi_{**}X$ for such X. Then we reproduce Eisen's calculation [**Eis87**] of $v_{n-1}^{-1}\operatorname{Ext}_{Y(n)}^{**}(\mathbf{F}_2,\mathbf{F}_2)$, where $Y(n)$ is the quotient Hopf algebra

$$Y(n) = \mathbf{F}_2[\xi_n, \xi_{n+1}, \ldots].$$

Finally, we discuss Mahowald's conjectured calculation [**Mah70**] at the prime 2 of $v_1^{-1}\pi_{**}M$, where M is an injective resolution of the homology of the mod 2 Moore spectrum. We note that the spectral sequence presented here will converge, and we have guesses for the differentials, but we do not know how to verify the guesses.

The main result of this chapter, the convergence of the localized spectral sequence, was originally proved by Margolis at the prime 2 in unpublished notes [**Mar**]; his approach was also presented in the author's Ph.D. thesis [**Pal91**], as was much of the analysis of the comodule structure of $A(n)$ in Sections 4.3–4.5. We have taken a different approach to convergence here, which was suggested by Hopkins.

4.1. The q_n-based Adams spectral sequence

We will use the following notation throughout this chapter.

NOTATION 4.1.1. We first introduced the spectra P_t^s and Q_n, as well as their connective covers p_t^s and q_n, in Section 2.2.

(a) When $p = 2$, we follow convention and write Q_n for the field spectrum P_{n+1}^0. Similarly, we write q_n for the connective version, the ring spectrum p_{n+1}^0. We computed the coefficients of these spectra in Proposition 2.2.3 and Corollary 2.2.5 and found that $(q_n)_{**} = \mathbf{F}_2[v_n]$ and $(Q_n)_{**} = \mathbf{F}_2[v_n^{\pm 1}]$; again, we are following convention and writing v_n for $h_{n+1,0}$. Recall that $|v_n| = (1, 2^{n+1} - 1)$.

(b) At odd primes, q_n, Q_n, and v_n are as usual. These have the same coefficient rings as at the prime 2, with $|v_n| = (1, 2p^n - 1)$.

(c) Let a_n denote the slope of Q_n, as defined in Notation 2.2.8:
$$a_n = \begin{cases} |\xi_{n+1}| = 2^{n+1} - 1, & \text{when } p = 2, \\ |\tau_n| = 2p^n - 1, & \text{when } p \text{ is odd.} \end{cases}$$

Regardless of the prime, the element v_n has bidegree $|v_n| = (1, a_n)$.

(d) Define the Hopf algebra E to be
$$E = \begin{cases} \mathbf{F}_2[\xi_{n+1}]/(\xi_{n+1}^2), & \text{when } p = 2, \\ E[\tau_n], & \text{when } p \text{ is odd.} \end{cases}$$

E is a quotient Hopf algebra of A, and indeed of any quotient A' of A in which ξ_{n+1} (respectively, τ_n) is nonzero. Regardless of the prime, $q_n = HE$.

We consider the q_n-based Adams spectral sequence. Ideally, we would like q_n to satisfy Condition 1.4.1: q_n should be a nice ring spectrum, q_n should be flat (i.e., $(q_n)_{**}(q_n)$ should be flat as a left module over $(q_n)_{**}$), and q_n should have good connectivity properties. q_n is defined to be HE, so it is a commutative and associative ring spectrum, by Proposition 1.2.8. We computed its coefficient ring in Proposition 2.2.3 (see also Notation 4.1.1), so we know that it satisfies the proper connectivity requirements. Because of this, the Adams spectral sequence converges to the homotopy of the "q_n-completion" of X. By Proposition 1.4.3, every connective spectrum is q_n-complete.

PROPOSITION 4.1.2. *Let X be a connective spectrum. Then the q_n-based Adams spectral sequence abutting to $\pi_{**}X$ converges.*

The next issue is whether q_n is flat, which would allow us to identify the E_2-term of the spectral sequence as an Ext group. It turns out that it isn't—see Corollary 4.6.2—so we do not have a formula for the E_2-term. Since we only use the q_n-based Adams spectral sequence as a tool to study the Q_n-based one, we do not need to know the E_2-term; hence this is not a big problem.

REMARK 4.1.3. If one works over a quotient A' of A in which ξ_{n+1} (respectively, τ_n) is nonzero, then the analogue of Proposition 4.1.2 holds. By "working over a quotient A' of A," we mean replacing the spectrum X with $HA' \wedge X$, as in

Theorem 2.3.6 (and see Remarks 5.1.4 and 5.1.7 for other examples). The resulting spectral sequence will converge to $\pi_{**}(HA' \wedge X) = HA'_{**}X$. If X is connective, so is $HA' \wedge X$, so Proposition 4.1.2 holds exactly as stated. As we progress through the chapter, we will make similar comments, most of which will have a bit more substance.

4.2. The Q_n-based Adams spectral sequence

In this section, we invert the element v_n in the q_n-based Adams spectral sequence converging to $\pi_{**}X$, in the hopes of computing $v_n^{-1}\pi_{**}X$. Of course, for $v_n^{-1}\pi_{**}X$ to make sense, some power of v_n must act on $\pi_{**}X$. The easiest way to arrange this is for the spectrum X under consideration to have an appropriate sort of self-map. Here is a special case of Definition 2.4.2; see also Definition 6.2.1.

DEFINITION 4.2.1. Fix a spectrum X and an integer $n \geq 0$. A self-map $f \in [X, X]_{**}$ is a v_n-*map* if for some j,
$$\eta_{Q_n} \wedge f = v_n^j \wedge 1_X$$
as maps $X \to Q_n \wedge X$. We write $v_n^{-1}X$ for the telescope
$$\varinjlim(X \xrightarrow{f} X \xrightarrow{f} X \xrightarrow{f} \cdots).$$

This is equivalent to requiring that $1_{Q_n} \wedge f = v_n^j \wedge 1_X$ in $[Q_n \wedge X, Q_n \wedge X]_{**}$—see Remark 6.2.2. If X is finite, it also implies that $\eta_{q_n} \wedge f^i = v_n^{ij} \wedge 1_X$ when $i \gg 0$.

REMARK 4.2.2. (a) $v_n \colon q_n \to q_n$ is an example of a v_n-map, and in this case, $Q_n = v_n^{-1} q_n$.
(b) In general, the telescope $\varinjlim(X \xrightarrow{f} X \xrightarrow{f} \cdots)$ may depend on the map f, so the notation $v_n^{-1}X$ is perhaps not ideal.
(c) One can use Theorem 2.4.3 to produce a v_n-map on any finite spectrum of type a_n, and any two such v_n-maps agree up to a power—see the statement and proof of Proposition 2.5.2(d). Hence in this case, $v_n^{-1}X$ is independent of the v_n-map. We focus on this situation below.
(d) Since Q_n is a field spectrum, then the Künneth formula says that if $f \colon X \to X$ is a v_n-map, then so is $1_Z \wedge f$ for any spectrum Z.

Suppose that X is a spectrum with a v_n-map. Then we claim that inverting v_n in the q_n-based Adams spectral sequence for $\pi_{**}X$ gives the Q_n-based Adams spectral sequence abutting to $\pi_{**}(v_n^{-1}X)$. Q_n is a commutative associative ring spectrum, and in fact a field spectrum, hence flat, so we can compute the E_2-term of this spectral sequence. Convergence is an issue, though, because Q_n is not connective. To summarize:

THEOREM 4.2.3. *Suppose that X is a connective spectrum with a v_n-map.*
(a) *The Q_n-based Adams spectral sequence abutting to $\pi_{**}(v_n^{-1}X)$ may be obtained from the q_n-based Adams spectral sequence converging to $\pi_{**}X$ by inverting some power of v_n at the E_2-term.*
(b) *The E_2-term of the Q_n-based Adams spectral sequence for $\pi_{**}(v_n^{-1}X)$ is*
$$E_2^{***} \cong \mathrm{Ext}^{***}_{(Q_n)_{**}(Q_n)}((Q_n)_{**}, (Q_n)_{**}X).$$
(c) *In this spectral sequence, multiplication by v_n increases degree by $(0, a_n - 1, 1)$.*
(d) *The differential d_r increases degree by $(r, r-1, 1-r)$.*

PROOF. Part (a): It suffices to show that given a q_n-Adams tower for X, inverting v_n turns it into a Q_n-Adams tower for $v_n^{-1}X$. The first section of Miller's paper [**Mil81**] is a good reference for the construction of Adams spectral sequences; it tells us that a q_n-Adams tower for X is a diagram of the form

(4.2.4) $$X \to K_0 X \to \Sigma^{-1,0} K_1 X \to \Sigma^{-2,0} K_2 X \to \cdots,$$

in which

(i) Each $K_s X$ is q_n-*injective*, meaning that $K_s X$ is a retract of a spectrum of the form $q_n \wedge Y$ for some Y, and

(ii) The sequence (4.2.4) is q_n-*exact*, meaning that applying $[-, J]$ gives an exact sequence for every q_n-injective spectrum J.

In fact, according to the construction of the Adams spectral sequence given in Section 1.4, we may assume that $K_s X$ is of the form $K_s X = q_n \wedge Z_s \wedge X$ for some spectrum Z_s. (Z_s is actually $\overline{q_n}^{\wedge s}$, where $\overline{q_n}$ is the fiber of the unit map of q_n, but we do not use this here.)

Let $f \colon X \to X$ denote the v_n-map of X; then $1_{Z_s} \wedge f$ is a v_n-map on $Z_s \wedge X$. By the remarks following Definition 4.2.1, after replacing f by a suitable power, then the two self-maps $v_n^j \wedge 1_{Z_s} \wedge 1_X$ and $1_{q_n} \wedge 1_{Z_s} \wedge f$ of $q_n \wedge Z_s \wedge X$ agree on homotopy; hence they have the same telescope. Thus if we invert $1 \wedge 1 \wedge f$ in each term of the tower, we get

$$\begin{array}{ccccccc} v_n^{-1} X & \longrightarrow & q_n \wedge Z_0 \wedge v_n^{-1} X & \longrightarrow & \Sigma^{-1,0} q_n \wedge Z_1 \wedge v_n^{-1} X & \longrightarrow & \cdots . \\ & & \| & & \| & & \\ & & Q_n \wedge Z_0 \wedge v_n^{-1} X & & \Sigma^{-1,0} Q_n \wedge Z_1 \wedge v_n^{-1} X & & \\ & & \| & & \| & & \\ & & Q_n \wedge Z_0 \wedge X & & \Sigma^{-1,0} Q_n \wedge Z_1 \wedge X & & \end{array}$$

Certainly, every term in this tower is Q_n-injective. To show that the tower is Q_n-exact, we apply $[-, Q_n \wedge Y]$. We claim that

(4.2.5) $$[Q_n \wedge Z_s \wedge X, Q_n \wedge Y] = [q_n \wedge Z_s \wedge X, Q_n \wedge Y] = v_n^{-1}[q_n \wedge Z_s \wedge X, q_n \wedge Y].$$

Given this, since localization is exact, then Q_n-exactness for the localized tower follows from q_n-exactness of the original tower. The first equality in Equation (4.2.5) follows because the fiber of

$$q_n \wedge Z_s \wedge X \to Q_n \wedge Z_s \wedge X$$

is killed by inverting v_n, so there are no maps from that fiber to Q_n, on which v_n is an equivalence. The second equality follows because $Q_n = v_n^{-1} q_n$.

Part (b): Since Q_n is flat, we can compute the E_2-term as an Ext group. Since X is connective, then $(Q_n)_{**} X = (Q_n)_{**}(v_n^{-1} X)$, so the E_2-term for the spectral sequence abutting to $\pi_{**}(v_n^{-1} X)$ is as given.

Part (c): We check this for the q_n-based Adams spectral sequence. Since v_n is in the coefficient ring $(q_n)_{**}$, then it has filtration zero in the spectral sequence. For it to act correctly in the abutment $\pi_{**} X$, then it must act with the indicated degree.

Part (d): This is true for any Adams spectral sequence. See Theorem 1.4.2. □

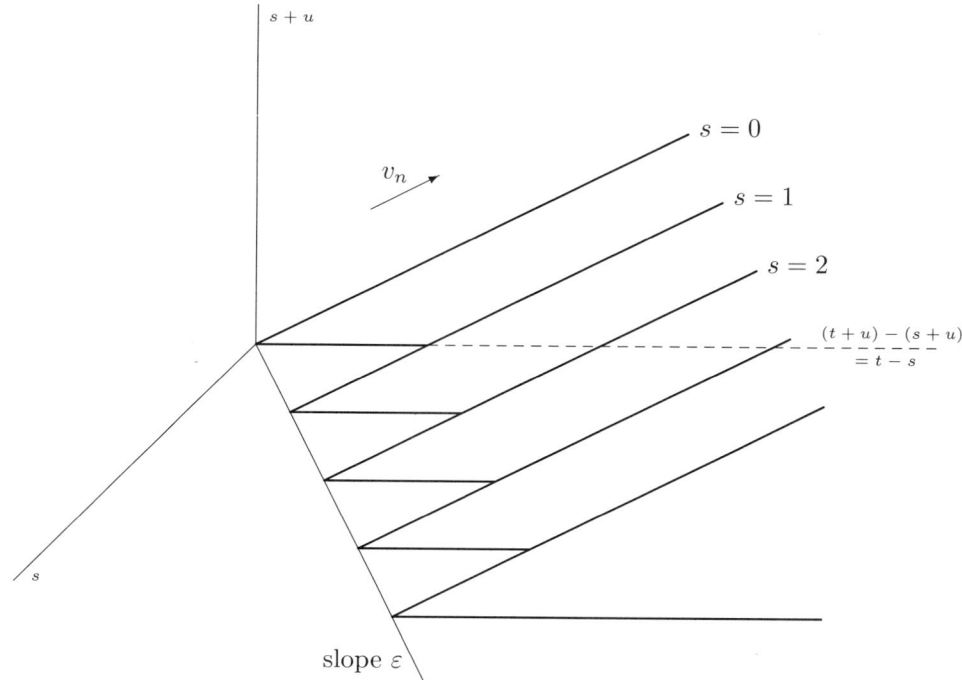

FIGURE 4.2.A. Vanishing plane in Theorem 4.2.6, with $b = 0$. The E_r-term is zero above the plane containing the lines labeled "$s = 0$," "$s = 1$," etc. The tridegrees with fixed $s+u$ and $t-s$ coordinates, with s increasing, form a composition series for $\pi_{s+u,t+u}$; we have chosen axes so that the bidegrees with the filtration degree s fixed are drawn with the usual Adams spectral sequence grading—see Remark 1.1.2. As indicated, multiplication by v_n moves parallel to the $(s+u, t-s)$-plane, at a slope of $\frac{1}{a_n-1}$. With this choice for the axes, d_r goes up one, left one, and increases s by r.

As mentioned above, convergence of the Q_n-based Adams spectral sequence is not guaranteed. Here is our first result about convergence; we use it to prove Theorem 4.3.1 below. Recall from Notation 4.1.1 that a_n is the slope of Q_n.

THEOREM 4.2.6. *Fix an integer $n \geq 0$. Let X be a connective spectrum with a v_n-map. Consider the Q_n-based Adams spectral sequence abutting to $v_n^{-1}\pi_{**}X$. If for some numbers r, b, and $\varepsilon > 0$, the E_r-term of the q_n-based Adams spectral sequence for X has a vanishing plane of the form*

(4.2.7) $$E_r^{s,t,u} = 0 \text{ when } s \geq \frac{-a_n}{\varepsilon}(s+u) + \frac{1}{\varepsilon}(t+u) + \frac{b}{\varepsilon},$$

then the Q_n-based spectral sequence converges.

We will only apply this result with $\varepsilon = 1$, but the theorem is just as easy to prove with a general ε.

See Figure 4.2.A for a picture of the vanishing plane.

PROOF. Consider the q_n-based Adams spectral sequence for $\pi_{**}X$, which converges because X is connective. Since X has a v_n-map, we can invert this map to get the Q_n-based Adams spectral abutting to $v_n^{-1}\pi_{**}X$. Since the v_n-map corresponds to multiplication by some power of v_n, say v_n^ℓ, then the q_n-based spectral sequence is one of $\mathbf{F}_p[v_n^\ell]$-modules, and the Q_n-based spectral sequence is obtained by inverting v_n^ℓ. When we have inverted v_n^ℓ, there are two ways the resulting spectral sequence can fail to converge to $v_n^{-1}\pi_{**}X$: differentials and extensions.

We will call a class x in the q_n-spectral sequence v_n-*periodic* if $(v_n^\ell)^j x \neq 0$ for all j. If a class is not v_n-periodic, we say that it is v_n-*torsion*.

Since the q_n-spectral sequence is one of $\mathbf{F}_p[v_n^\ell]$-modules, the only allowable differentials go from v_n-periodic classes to other v_n-periodic classes, or from v_n-periodic classes to v_n-torsion classes. The first sort of differentials don't cause any problems with convergence; the second sort only cause problems if an infinite sequence of them conspires to kill off a v_n-tower; none of these differentials would be present in the localized spectral sequence, so at E_∞ of the localized spectral sequence, there would be a superfluous v_n-tower.

The second possible problem is extensions: if, at E_∞ of the q_n-spectral sequence, an infinite collection of v_n-torsion families is connected by extensions to form a single v_n-tower, this tower will not be present in the localized spectral sequence, so the localized spectral sequence will give the wrong answer: it won't have enough v_n-towers.

We claim that the vanishing plane in (4.2.7) will prevent both of these problems. Suppose that we have the vanishing plane, as given, at the E_{r_0}-term, and hence at the E_r-term for all $r \geq r_0$. Note that since $|v_n| = (0, a_n - 1, 1)$ in the spectral sequence, the plane is parallel to the action by v_n. Given a class $y \in E_r^{i,j,k}$, then $d_r y$ has degree
$$|d_r y| = (i+r, j+r-1, k+1-r).$$
We check the vanishing plane condition: after multiplying through by $\varepsilon > 0$, we have
$$\varepsilon(i+r) \overset{?}{\geq} -a_n(i+k+1) + (j+k) + b.$$
Since ε is positive, then for all sufficiently large r this inequality will hold; hence $d_r y$ is above the vanishing plane for all $r \gg 0$, and hence zero. In other words, each class in the spectral sequence can only support a differential d_r if r is small enough. Since the vanishing plane is parallel to the v_n-action, then the class y and all the classes $\{v_n^d y\}$ are affected the same way by the plane's presence.

Similarly, along any given line in the abutment $\pi_{s+u,t+u}$ of slope a_n, the E_∞-term is nonzero in only finitely many filtrations above that line; hence there can not be infinitely many extensions, so it is impossible to produce a v_n-tower from v_n-torsion families. □

Notice that if one can find a spectrum X which satisfies the vanishing plane condition at some E_r, then every spectrum in the thick subcategory generated by X also satisfies it, perhaps at a different term of the spectral sequence, by Theorem 1.4.4.

REMARK 4.2.8. If one works over a quotient A' of A in which ξ_{n+1} (respectively, τ_n) is nonzero, one has similar results: the Q_n-based spectral sequence is

4.3. $A(n)$ as an A-comodule

Here is our goal for the next three sections:

THEOREM 4.3.1. *Fix $n \geq 0$. The Q_n-based Adams spectral sequence abutting to $v_n^{-1}\pi_{**}X$ converges for all finite X which are $\bigvee_{d<a_n} Z(d)_{**}$-acyclic.*

Notice that by Theorem 2.4.3, every finite $\bigvee_{d<a_n} Z(d)_{**}$-acyclic spectrum has a v_n-map, so it makes sense to discuss $v_n^{-1}\pi_{**}X$.

To prove the theorem, we will use a finite spectrum called $\frac{1}{2}A(n)$. Recall that we defined the quotient Hopf algebras $A(n)$ of A in Example 2.1.4:

$$A(n) = \begin{cases} A/(\xi_1^{2^{n+1}}, \xi_2^{2^n}, \ldots, \xi_{n+1}^2, \xi_{n+2}, \xi_{n+3}, \ldots), & p=2, \\ A/(\xi_1^{p^n}, \xi_2^{p^{n-1}}, \ldots, \xi_n^p, \xi_{n+1}, \xi_{n+2}, \ldots; \tau_{n+1}, \tau_{n+2}, \ldots), & p \text{ odd}. \end{cases}$$

By a theorem of Mitchell [**Mit85**], and independently, Smith [**Smi**], $A(n)$ admits a self-dual A-comodule structure extending its left $A(n)$-comodule structure. Let $\frac{1}{2}A(n)$ denote the sub-vector space of $A(n)$ defined by

$$\frac{1}{2}A(n) = A(n) \square_E \mathbf{F}_p,$$

with E as in Notation 4.1.1. $\frac{1}{2}A(n)$ is certainly an $A(n)$-subcomodule of $A(n)$; we note in Corollary 4.3.8 below that it in fact inherits an A-comodule structure from $A(n)$. We write $\frac{1}{2}A(n)$ for both the comodule and its injective resolution, as long as it is clear from context which we intend.

We can now prove the main theorem, modulo a few details.

PROOF OF THEOREM 4.3.1. Theorem 4.4.1 below says that $\frac{1}{2}A(n)$ satisfies the vanishing plane condition (4.2.7) at the E_2-term of the q_n-based Adams spectral sequence. Theorem 4.5.1 says that $\frac{1}{2}A(n)$ generates the thick subcategory of all finite $\bigvee_{d<a_n} Z(d)_{**}$-acyclics. By the genericity of vanishing planes, Theorem 1.4.4, we conclude that every $\bigvee_{d<a_n} Z(d)_{**}$-acyclic satisfies the vanishing plane condition. By Theorem 4.2.6, the Q_n-based Adams spectral sequence converges for these spectra, as desired. □

REMARK 4.3.2. If one works over a quotient A' of A in which ξ_{n+1} (respectively, τ_n) is nonzero, then $HA' \wedge \frac{1}{2}A(n)$ has the same properties as $\frac{1}{2}A(n)$, hence the analogue of Theorem 4.3.1 holds in that context. Note that verifying this does not require reproving all of the results in the next few sections, but rather smashing those results with HA'.

For the rest of this section, we discuss some of the basic properties of the object $\frac{1}{2}A(n)$. In Section 4.4 we prove Theorem 4.4.1: the q_n-based Adams spectral sequence for $\frac{1}{2}A(n)$ has a nice vanishing plane at the E_2-term. This requires the construction of a specific q_n-Adams tower for $\frac{1}{2}A(n)$. In Section 4.5 we prove Theorem 4.5.1: $\frac{1}{2}A(n)$ generates the thick subcategory of all finite $\bigvee_{d<a_n} Z(d)_{**}$-acyclics. Parts of this are straightforward, but one piece requires a careful analysis of Mitchell's and Smith's A-comodule structure on $A(n)$. We do this careful analysis in Subsection 4.5.1.

$\frac{1}{2}A(n)$ is a sub-vector space, and we claim an A-subcomodule, of $A(n)$; to verify this, we need to examine some features of the A-comodule structure on $A(n)$. First, we establish some notation. The main difference between $A(n)$ at the prime 2 and $A(n)$ at odd primes is, obviously, the τ_i's. It is sometimes useful to deal with the Hopf algebra generated by just the ξ_i's at all primes simultaneously; the following gives us notation to do that.

DEFINITION 4.3.3. Let $P(n)$ be this quotient Hopf algebra of A:
$$P(n) = \mathbf{F}_p[\xi_1, \xi_2, \xi_3, \dots]/(\xi_1^{p^{n+1}}, \xi_2^{p^n}, \dots, \xi_n^{p^2}, \xi_{n+1}^p, \xi_{n+2}, \xi_{n+3}, \dots).$$

So when $p = 2$, $A(n) = P(n)$. Mitchell [**Mit85**] and Smith [**Smi**] also produce an A-comodule structure on $P(n)$ extending its $P(n)$-comodule structure.

REMARK 4.3.4. Mitchell's and Smith's constructions of comodule structures on $A(n)$ and $P(n)$ are somewhat different—Mitchell uses $GL_m(\mathbf{F}_2)$-invariants, and Smith uses idempotents in the group algebra $\mathbf{F}_2[\Sigma_r]$, for appropriate numbers m and r—but the main result of [**Mit86**] is that these two constructions result in the same A-comodule structure. Note also that [**Mit86**] is a good reference for Smith's construction; the construction is also described in [**Rav92**].

There may be many possible A-comodule structures on $A(n)$ and $P(n)$—at the prime 2, one can see that there are four different comodule structures on $P(1) = A(1)$ (depending on the possible coactions of ξ_1^4), and hundreds of different comodule structures on $P(2) = A(2)$ (see [**Rot77**]). We will use Mitchell's and Smith's comodule structure on $P(n)$; when p is odd, we will use another comodule structure on $A(n)$, which we now describe.

DEFINITION 4.3.5. For p an odd prime and $n \geq 0$, let $V(n)$ denote the Hopf algebra
$$V(n) = \Lambda[\tau_0, \dots, \tau_n].$$
Let $V(-1) = \mathbf{F}_p$.

Then $V(n)$ is a quotient Hopf algebra of A and of $A(n)$. $V(-1)$ is obviously an A-comodule, and it turns out that the same is true for all of the $V(n)$'s.

LEMMA 4.3.6. *Fix an odd prime p and $n \geq 0$.*
(a) *$V(n)$ has an A-comodule structure extending its $V(n)$-comodule structure.*
(b) *There is a short exact sequence of A-comodules*
$$0 \to V(n-1) \to V(n) \to s^{|\tau_n|}V(n-1) \to 0.$$
Here, s is the internal suspension functor on the comodule category—see Section 0.1.
(c) *If M is an A-comodule which is cofree of rank one as a $P(n-1)$-comodule, and if N is an A-comodule which is cofree of rank one as a $V(n)$-comodule, then $M \otimes N$ is an A-comodule which is cofree of rank one as an $A(n)$-comodule.*

PROOF. Part (a): The coproduct on τ_k,
$$\Delta \tau_k = \sum_{i=0}^{k} \xi_{k-i}^{p^i} \otimes \tau_i + \tau_k \otimes 1,$$
extended multiplicatively, defines an A-comodule structure map $V(n) \to A \otimes V(n)$.
Part (b): This is clear, given part (a).

4.3. $A(n)$ AS AN A-COMODULE

Part (c): An $A(n)$-comodule L is cofree over $A(n)$ if and only if it is cofree over $P(n-1)$ and $V(n)$. An appeal to Lemma 0.4.3 finishes the proof. □

NOTATION 4.3.7. It is important which A-comodule structure we use on these Hopf algebras, so we fix Mitchell's and Smith's A-comodule structures on $P(n)$, the above comodule structure on $V(n)$, and the structure $P(n-1) \otimes V(n)$ on $A(n)$. We will use $A(n)$ and $P(n)$ to denote the Hopf algebras, the comodules with the given structure, and their injective resolutions, depending on the context. For the remainder of this section, we will use the notation $A(n)$ only at odd primes.

COROLLARY 4.3.8. *There is an A-comodule structure on $\frac{1}{2}A(n)$ which extends its $A(n)$-comodule structure.*

PROOF. When $p = 2$, this is part of [**Mit85**, Corollary 3.14(b)]. When p is odd, then $P(n-1) \otimes V(n-1)$ is an A-comodule which is isomorphic, as an $A(n)$-comodule, to $\frac{1}{2}A(n)$. □

REMARK 4.3.9. We should point out that the quotient Hopf algebra structure and the A-comodule structure on $A(n)$ do not interact as well as they might; the same goes for $P(n)$. To start with, while $A(n)$ is a quotient Hopf algebra of A, the quotient map $\pi\colon A \twoheadrightarrow A(n)$ is not a map of A-comodules. For example, chase the primitive element $\xi_1^{p^{n+1}}$ around the following diagram:

$$\begin{array}{ccc} A & \xrightarrow{\pi} & A(n) \\ \Delta \downarrow & & \downarrow \psi \\ A \otimes A & \xrightarrow{1 \otimes \pi} & A \otimes A(n). \end{array}$$

Going one way, $\psi \circ \pi(\xi_1^{p^{n+1}}) = 0$; going the other way, $(1 \otimes \pi) \circ \Delta(\xi_1^{p^{n+1}}) = \xi_1^{p^{n+1}} \otimes 1$. Of course, since A is cofree of rank one (Definition 0.4.1), then there is a unique A-comodule map $i\colon A(n) \to A$ which sends 1 to 1. As an $A(n)$-comodule, $A(n)$ is cofree of rank one, so there is a unique $A(n)$-comodule map $A(n) \to A(n)$ so that $1 \mapsto 1$, namely the identity map. Since the A-comodule structure on $A(n)$ extends its $A(n)$-comodule structure, the following diagram commutes:

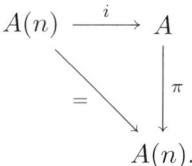

Hence the A-comodule map $i\colon A(n) \to A$ is one-to-one. Similar comments apply to $P(n)$, so there is a one-to-one A-comodule map $j\colon P(n) \to A$. Unfortunately, the inclusions i and j are not, in general, coalgebra maps. Consider this diagram:

$$\begin{array}{ccc} P(n) & \xrightarrow{i} & A \\ \Delta \downarrow & & \downarrow \Delta \\ P(n) \otimes P(n) & \xrightarrow{i \otimes i} & A \otimes A. \end{array}$$

Let $p = 2$ and $n = 1$ and consider the element $\xi_1^3 \xi_2 \in P(1)$. We have $i(\xi_1^3 \xi_2) = \xi_1^3 \xi_2 + c\xi_1^6$ for some scalar c which depends on the choice of A-comodule structure on $P(1)$. Regardless of what c is, one summand in $\Delta \circ i(\xi_1^3\xi_2)$ is $\xi_1^5 \otimes \xi_1$, whereas $(i \otimes i) \circ \Delta(\xi_1^3\xi_2)$ has no such term. (In module language, this corresponds to the fact that no matter which A^*-module structure one puts on $P(1)^*$, the Adem relations force a nonzero action by Sq^4; therefore the A^*-module quotient map $A^* \twoheadrightarrow P(1)^*$ is not an algebra map.)

4.4. $\frac{1}{2}A(n)$ satisfies the vanishing plane condition

Recall that the number a_n, defined in Notation 4.1.1, is the slope of Q_n. Our goal here is to prove the following.

THEOREM 4.4.1. *Fix $n \geq 0$. The q_n-based Adams spectral spectral sequence converging to $\pi_{s+u,t+u}(\frac{1}{2}A(n))$ satisfies:*

$$E_2^{s,t,u} = 0 \text{ when } s \geq -a_n(s+u) + (t+u).$$

In other words, it satisfies the vanishing plane condition (4.2.7) with $\varepsilon = 1$ and $b = 0$.

The philosophy behind the proof is that there is a nice relationship between $(Q_n)_{**}(Q_n)$ and $\frac{1}{2}A(n)$: we show in Proposition 4.6.3 below that

$$(Q_n)_{**}(Q_n) \cong \mathbf{F}_p[v_n^{\pm 1}] \otimes H(A \,\square_E \mathbf{F}_p, Q_n).$$

Here, E is as in Notation 4.1.1. We compute the Hopf algebra $\Gamma = H(A \,\square_E \mathbf{F}_p, Q_n)$ in Proposition 4.6.1; from that computation, we can see that $\frac{1}{2}A(n)$ is a quotient Hopf algebra of Γ, and the two agree in dimensions less than $a_n + 1$. $\frac{1}{2}A(n)$ also has a Γ-comodule structure. As a result, $\frac{1}{2}A(n)$ has a Γ-resolution in which the terms increase rapidly in connectivity; in fact, they increase by at least $a_n + 1$ at each stage.

The statements in the above paragraph can be translated directly into a vanishing plane for the E_2-term of the Q_n-based Adams spectral sequence. We want to look at the q_n-based spectral sequence, though, so we give the same argument, but in somewhat different language. The following proof does not use any results from Section 4.6.

PROOF. We would like to show the spectrum $\frac{1}{2}A(n)$ satisfies the vanishing plane condition (4.2.7) at the E_2-term of the q_n-based Adams spectral sequence. Since we do not have a formula for that E_2-term in general, we analyze it here by using a specific Adams tower; for this specific Adams tower, we in fact show that we have the desired vanishing plane at E_1.

As $A(n)$-comodules, we have $\frac{1}{2}A(n) = A(n) \,\square_E \mathbf{F}_p$; therefore, as A-comodules we have $(A \,\square_{A(n)} \mathbf{F}_p) \otimes \frac{1}{2}A(n) \cong A \,\square_E \mathbf{F}_p$, using the shearing isomorphism, Lemma 1.2.5. As spectra, this translates into the equality

$$HA(n) \wedge \frac{1}{2}A(n) = HE = q_n.$$

We use this fact to build a q_n-Adams tower for $\frac{1}{2}A(n)$.

LEMMA 4.4.2. *The canonical $HA(n)$-Adams tower for $\frac{1}{2}A(n)$ is a q_n-Adams tower for $\frac{1}{2}A(n)$.*

We prove this below.

4.4. $\frac{1}{2}A(n)$ SATISFIES THE VANISHING PLANE CONDITION

Recall from Section 1.4 that the canonical $HA(n)$-based Adams tower for $X = \frac{1}{2}A(n)$ is of the form

(4.4.3)
$$\begin{array}{ccccccc} X & = & F_0X & \longleftarrow & F_1X & \longleftarrow & F_2X & \longleftarrow & \cdots, \\ & & \downarrow & & \downarrow & & \downarrow & & \\ & & K_0X & & K_1X & & K_2X & & \end{array}$$

where
$$F_sX = \overline{HA(n)}^{\wedge s} \wedge X,$$
$$\overline{HA(n)} = \text{fiber } (S^0 \to HA(n)),$$
$$K_sX = HA(n) \wedge \overline{HA(n)}^{\wedge s} \wedge X.$$

The comodule $A \,\square_{A(n)} \mathbf{F}_p$ is an $\mathbf{F}_p$-algebra; let $\overline{A \,\square_{A(n)} \mathbf{F}_p}$ denote the cokernel of the unit map $\eta \colon \mathbf{F}_p \to A \,\square_{A(n)} \mathbf{F}_p$. This unit map is an A-comodule map, so $\overline{A \,\square_{A(n)} \mathbf{F}_p}$ is an A-comodule.

Lemma 4.4.2 says that the spectral sequence obtained by applying $\pi_{s+u,t+u}$ to the tower in (4.4.3) agrees with the q_n-based Adams spectral sequence starting with the E_2-term. Our next goal is to compute the E_1-term obtained from (4.4.3). We use the following lemma here, and also to prove Lemma 4.4.2.

LEMMA 4.4.4. *Let $X = \frac{1}{2}A(n)$.*
(a) $K_sX = q_n \wedge \overline{HA(n)}^{\wedge s}$.
(b) $\Sigma^{-1,0}\overline{HA(n)}$ *is an injective resolution of the comodule* $\overline{A \,\square_{A(n)} \mathbf{F}_p}$.

We prove this below. By the lemma, we have
$$E_1^{s,t,u} = \pi_{s+u,t+u}(K_sX)$$
$$= \pi_{s+u,t+u}(q_n \wedge \overline{HA(n)}^{\wedge s})$$
$$= (q_n)_{s+u,t+u}(\overline{HA(n)}^{\wedge s})$$
$$= (q_n)_{u,t+u}(\Sigma^{-s,0}\overline{HA(n)}^{\wedge s})$$
$$= \text{Ext}_E^{u,t+u}(\mathbf{F}_p, \overline{A \,\square_{A(n)} \mathbf{F}_p}^{\otimes s}).$$

For any E-comodule M, $\text{Ext}_E^{u,t+u}(\mathbf{F}_p, M)$ is a direct sum of copies of $\mathbf{F}_p$, which occur when $u = 0$, and copies of $\mathbf{F}_p[v_n]$, which start when $u = 0$—see Proposition 0.5.10. Recall that in this setting, the (s,t,u)-degree of v_n is $(0, a_n - 1, 1)$; hence if the comodule M has bottom cell in degree m, this Ext group is zero when
$$a_n u \geq (t + u) + m.$$

Since $A \twoheadrightarrow A(n)$ is an isomorphism in degrees less than $a_n + 1$, then the bottom class in $\overline{A \,\square_{A(n)} \mathbf{F}_p}$ is in degree at least $a_n + 1$, so $\overline{A \,\square_{A(n)} \mathbf{F}_p}^{\otimes s}$ starts in degree at least $(a_n + 1)s$. Hence
$$\text{Ext}_E^{u,t+u}(\mathbf{F}_p, \overline{A \,\square_{A(n)} \mathbf{F}_p}^{\otimes s}) = 0$$
when $(a_n + 1)s + a_n u \geq (t + u)$, or equivalently, when
$$s \geq -a_n(s + u) + (t + u),$$
as claimed. $\square$

PROOF OF LEMMA 4.4.2. Turning the tower (4.4.3) around, we have
$$(4.4.5) \qquad X \to K_0 X \to \Sigma^{-1,0} K_1 X \to \Sigma^{-2,0} K_2 X \to \cdots.$$
We use the criteria from [Mil81], reproduced in the proof of Theorem 4.2.3(a), to check that (4.4.3) is a q_n-based Adams tower for $X = \frac{1}{2}A(n)$. It is clear that every term in the tower is q_n-injective, since each $K_s X$ is already of the form $q_n \wedge Y$. It suffices to check q_n-exactness for spectra J of the form $J = q_n \wedge Y$. In other words, we want to prove exactness for the sequence
$$[\Sigma^{-1,0} K_{s+1} X, J] \xrightarrow{f_s^*} [K_s X, J] \xrightarrow{f_{s-1}^*} [\Sigma^{1,0} K_{s-1} X, J].$$
By Lemma 4.4.4, $K_s X$ is the $(s,0)$-suspension of an injective resolution of the comodule
$$(A \square_E \mathbf{F}_p) \otimes \overline{A \square_{A(n)} \mathbf{F}_p}^{\otimes s}.$$
Furthermore, the map $K_s X \to \Sigma^{-1,0} K_{s+1} X$ comes from tensoring the following composition of maps of comodules with $\frac{1}{2} A(n) \otimes \overline{A \square_{A(n)} \mathbf{F}_p}^{\otimes s}$:
$$A \square_{A(n)} \mathbf{F}_p \xrightarrow{g} \overline{A \square_{A(n)} \mathbf{F}_p} \xrightarrow{\eta \otimes 1} (A \square_{A(n)} \mathbf{F}_p) \otimes (\overline{A \square_{A(n)} \mathbf{F}_p}),$$
where $g \colon A \square_{A(n)} \mathbf{F}_p \to \overline{A \square_{A(n)} \mathbf{F}_p}$ is the quotient map and $\eta \colon \mathbf{F}_p \to A \square_{A(n)} \mathbf{F}_p$ is the unit map. Hence the entire sequence (4.4.5) may be obtained by taking injective resolutions of a sequence of comodules.

We point out that as an E-comodule, $A \square_{A(n)} \mathbf{F}_p$ splits via the maps g and η as $A \square_{A(n)} \mathbf{F}_p \cong_E \mathbf{F}_p \oplus \overline{A \square_{A(n)} \mathbf{F}_p}$. One can see this by first recalling from Corollary 0.5.5 that every finite-type E-comodule breaks into a direct sum of cofree comodules and trivial comodules, and then noting that $A \square_{A(n)} \mathbf{F}_p$ is zero in dimension a_n, so the class in dimension 0 must give a trivial summand, not a cofree one.

Given all of this, it is easy to prove q_n-exactness of (4.4.5) at $[K_s X, HE \wedge Y]$. By Corollary 1.2.7, this is the same as maps over E from $K_s X$ to Y:
$$[K_s X, HE \wedge Y] = [K_s X, Y]^E.$$
Over E, though, the map $K_{s-1} X \to K_s X$ is the composite of first projecting $K_{s-1} X$ onto a summand, and then including that as a summand of $K_s X$. The image of this map is clearly the kernel of $K_s X \to K_{s+1} X$. It follows that (4.4.5) is q_n-exact. □

PROOF OF LEMMA 4.4.4. Part (a): Since $X \wedge HA(n) = q_n$, then $K_s X = q_n \wedge \overline{HA(n)}^{\wedge s}$, as desired.

Part (b): S^0 is an injective resolution of $\mathbf{F}_p$, $HA(n)$ is an injective resolution of $A \square_{A(n)} \mathbf{F}_p$, and the unit map $S^0 \to HA(n)$ is induced by the monomorphism $\mathbf{F}_p \hookrightarrow A \square_{A(n)} \mathbf{F}_p$. Hence $\overline{HA(n)}$ is an injective resolution of the cokernel $\overline{A \square_{A(n)} \mathbf{F}_p} = A \square_{A(n)} \mathbf{F}_p / \mathbf{F}_p$, suspended: $\overline{HA(n)} = \Sigma^{1,0} I$, where I is an injective resolution of $\overline{A \square_{A(n)} \mathbf{F}_p}$. □

4.5. $\frac{1}{2}A(n)$ generates the expected thick subcategory

In this section, we examine the thick subcategory generated by the object $\frac{1}{2}A(n)$. (Throughout, we will use $\frac{1}{2}A(n)$ to denote both the comodule and its injective resolution, as long as the context makes it clear which we mean.) In particular, we prove the following.

THEOREM 4.5.1.   (a) $Z(d)_{**}\frac{1}{2}A(n)$ is zero if and only if $d < a_n$.
(b) For any finite spectrum W which is $Z(d)_{**}$-acyclic for all $d < a_n$, W is in thick($\frac{1}{2}A(n)$).

Let d_m be the slope preceding a_n,

$$d_m = a_n - 1 = \begin{cases} |\xi_n^2| & \text{when } p = 2, \\ |\xi_n| & \text{when } p \text{ is odd.} \end{cases}$$

Then Theorem 2.5.3(a) says that the spectrum $F(u_{d_1}^{j_1}, \ldots, u_{d_m}^{j_m})$ has the same properties. We will prove Theorem 4.5.1 the same way we proved that one: we will show that $\frac{1}{2}A(n)$ can be built as an iterated cofiber of well-understood self-maps of finite spectra.

We need some definitions and preliminary results before we start the proof. The Hopf algebra $V(k)$ was defined in Definition 4.3.5; we produced an A-comodule structure on it in Lemma 4.3.6.

DEFINITION 4.5.2.   (a) Suppose that p is odd. For any $k \geq 0$, define the vector space $A(n)_k$ to be $A(n)_k = P(n-1) \otimes V(k-1)$.
(b) For any prime p, for any $k \geq 1$, let $B(k)$ be this quotient Hopf algebra of A:

$$B(k) = \mathbf{F}_p[\xi_k, \xi_{k+1}, \xi_{k+2}, \ldots]/(\xi_i^p : i \geq k),$$

and let

$$P(n)_k = \ker\left(P(n) \xrightarrow{\overline{\psi}} JA \otimes P(n) \to JB(k) \otimes P(n)\right),$$

where $JA = \text{cok}(\mathbf{F}_p \to A)$ is the coaugmentation coideal of A, $JB(k)$ is the coaugmentation coideal of $B(k)$, and $\overline{\psi}$ is the reduced coaction map on $P(n)$. (See Definition 1.2.12(b) for "coaugmentation coideal.")

For example, $A(n)_{n+1} = A(n)$ and $A(n)_0 = P(n-1)$; more generally, if $k \leq n+1$, then each $A(n)_k$ is an A-subcomodule of $A(n)$. We show in Proposition 4.5.6 that each $P(n)_k$ is an A-subcomodule of $P(n)$. To start with, we have the following.

PROPOSITION 4.5.3. Fix $k > n+1$. For all primes, we have $P(n)_k = P(n)$. In particular, if $k > n+1$, then $P(n)_k$ is an A-comodule.

PROOF. This is due to Mitchell: when $p = 2$, [**Mit85**, Corollary 3.14(a)] is exactly the result that $P(n)_k = P(n)$ when $k > n+1$. When p is odd and $k > n+1$, the proof of Mitchell's result can be imitated to show that $P(n)_k = P(n)$. (Alternatively, this result says that ξ_k "coacts trivially" on $P(n)$ if $k > n+1$—there are no terms $\xi_k \otimes x$ in the A-coaction map on elements of $P(n)$. We describe Smith's construction of the A-comodule structure on $P(n)$ in Subsection 4.5.1; $P(n)$ is constructed as a summand of a tensor product of comodules on which ξ_k coacts trivially; as a result, ξ_k coacts trivially on the entire tensor product, and hence on $P(n)$.) □

COROLLARY 4.5.4. As a sub-vector space of $P(n)$ when $p = 2$ (respectively, of $A(n)$ when p is odd), we have

$$\frac{1}{2}A(n) = \begin{cases} P(n)_{n+1} & \text{when } p = 2, \\ A(n)_n & \text{when } p \text{ is odd.} \end{cases}$$

PROOF. At the prime 2, $\frac{1}{2}A(n)$ is defined to be $P(n)\,\square_E\mathbf{F}_2$, with E as in Notation 4.1.1. Since $P(n)_{n+2} = P(n)$, then

$$P(n)_{n+1} = \ker\Big(P(n) \to JE \otimes P(n)\Big)$$
$$= \mathbf{F}_2 \,\square_E P(n)$$
$$= P(n) \,\square_E \mathbf{F}_2,$$

the last equality holding because E is a conormal quotient of $P(n)$.

The p odd case is even easier. In this case, $\frac{1}{2}A(n) = A(n)\,\square_E \mathbf{F}_p$, and one can see from our comodule structure on $A(n) = P(n-1) \otimes V(n)$ that $\frac{1}{2}A(n) = P(n-1) \otimes V(n-1)$, as desired. □

In order to prove Theorem 4.5.1, we will need the following results. The Ext elements v_k, h_{k0}, and b_{k0} are defined in Remark 2.1.3.

PROPOSITION 4.5.5. *For each k with $0 \le k \le n$, there is a short exact sequence of A-comodules*

$$0 \to A(n)_k \to A(n)_{k+1} \to s^{|\tau_k|}A(n)_k \to 0.$$

The short exact sequence corresponds to $v_k \in \operatorname{Ext}^1_A(A(n)_k, A(n)_k)$.

PROOF. This follows from Lemma 4.3.6(b). □

We have a similar result for $P(n)$. The proof is more involved, so it will help to have a more precise statement.

PROPOSITION 4.5.6. *Fix k with $1 \le k \le n+1$.*

(a) *$P(n)_k$ is an A-subcomodule of $P(n)_{k+1}$, and it is spanned by the monomials in $A(n)$ in which each ξ_i, $k \le i \le n+1$, occurs to a power divisible by p.*

(b) *When $p = 2$, there is a short exact sequence of A-comodules*

$$0 \to P(n)_k \to P(n)_{k+1} \to s^{|\xi_k|}P(n)_k \to 0,$$

corresponding to $h_{k0} \in \operatorname{Ext}^1_A(P(n)_k, P(n)_k)$.

(c) *When p is odd, there is an A-comodule M and two short exact sequences of A-comodules*

$$0 \to P(n)_k \to P(n)_{k+1} \to s^{|\xi_k|}M \to 0,$$
$$0 \to M \to P(n)_{k+1} \to s^{|\xi_k^{p-1}|}P(n)_k \to 0,$$

corresponding to elements $\alpha \in \operatorname{Ext}^1_A(M, P(n)_k)$ and $\beta \in \operatorname{Ext}^1_A(P(n)_k, M)$, respectively, such that $\beta\alpha = b_{k0}$ in $\operatorname{Ext}^2_A(P(n)_k, P(n)_k)$.

PROOF. Part (a): We prove this by downward induction on k. The induction starts with $P(n)_{n+2} = P(n)$, which we know is an A-comodule with the advertised monomial basis. Given that $P(n)_{k+1}$ has the right basis, we will show that $P(n)_k$ is an A-subcomodule and that its basis is of the correct form.

By the definition 4.5.2 of $P(n)_k$, we have

$$P(n)_k = \ker\Big(P(n)_{k+1} \to JD[\xi_k] \otimes P(n)_{k+1}\Big),$$

where the map is induced by the coaction map. The coaugmentation coideal $JD[\xi_k]$ is equal to $\operatorname{Span}_{\mathbf{F}_p}(\xi_k, \ldots, \xi_k^{p-1})$, and by coassociativity, an element x has a term

$\xi_k^i \otimes y$ in its coproduct for some i with $0 < i < p$ if and only if it has a term $\xi_k \otimes z$. Hence

$$P(n)_k = \ker\left(P(n)_{k+1} \to \mathrm{Span}(\xi_k) \otimes P(n)_{k+1}\right)$$
$$= \ker\left(P(n)_{k+1} \to s^{|\xi_k|} P(n)_{k+1}\right),$$

where the map is defined by first applying the coaction map on $P(n)_k$, then projecting from A to $\mathrm{Span}(\xi_k)$. In other words, the map is multiplication by P_k^0, the element of A^* dual to ξ_k. We claim that this is an A-comodule map, and hence its kernel, $P(n)_k$, inherits an A-comodule structure from $P(n)_{k+1}$. We have to check that this diagram commutes:

$$\begin{array}{ccc} P(n)_{k+1} & \xrightarrow{P_k^0} & s^{|\xi_k|} P(n)_{k+1} \\ \psi \downarrow & & \downarrow \psi \\ A \otimes P(n)_{k+1} & \xrightarrow{1 \otimes P_k^0} & A \otimes s^{|\xi_k|} P(n)_{k+1}. \end{array}$$

Since $P(n)_{k+1}$ is an A-subcomodule of $P(n)$, and since we have formulas for the $P(n)$-coaction on $P(n)$, then we understand the $P(n)$-coaction on $P(n)_{k+1}$. In particular, the only way for a monomial $y \in P(n)_{k+1}$ to have a summand $\xi_k \otimes z$ in its coproduct is if ξ_k^j is a factor in y, where j is not divisible by p. In this case, multiplication by P_k^0 takes $\xi_k^j z$ to $\xi_k^{j-1} z$. Hence the kernel of P_k^0 is spanned by the monomials in $P(n)_{k+1}$ in which ξ_k occurs to a power divisible by p. Furthermore, we can see that if x is in the kernel of P_k^0, then it goes to zero each way around the diagram. Given a monomial y which is not in the kernel, then y is of the form $y = \xi_k z$ for some z, and $P_k^0(y) = \sigma z$, where we have written σz for the element

$$\sigma z = \xi_k \otimes z \in \mathrm{Span}(\xi_k) \otimes P(n)_{k+1}.$$

If the comultiplication map applied to z gives $\sum_i a_i \otimes z_i$, then by coassociativity, going each way around the diagram yields $\sum_i a_i \otimes (\sigma z_i)$. In other words, the diagram commutes for all monomials in $P(n)_{k+1}$. Since the monomials form a basis, we conclude that the diagram commutes.

We have just shown that $P_k^0 \colon P(n)_{k+1} \to P(n)_{k+1}$ is an A-comodule map; hence its kernel $P(n)_k$ is an A-subcomodule of $P(n)_{k+1}$. We also verified the monomial basis for $P(n)_k$. This completes the inductive step, and hence the proof of part (a).

Part (b): By part (a), $P(n)_k$ is an A-subcomodule of $P(n)_{k+1}$, so we have to check that the cokernel of the inclusion map is isomorphic to the appropriate suspension of $P(n)_k$. But we understand exactly what that cokernel is, by our description of the map P_k^0: for $x \in P(n)_k$, we have

$$P(n)_{k+1} \xrightarrow{P_k^0} s^{|\xi_k|} P(n)_{k+1}.$$
$$x \longmapsto 0$$
$$\xi_k x \longmapsto \sigma x$$

Since multiplication by P_k^0 is an A-comodule map, then the A-coaction on σx is σ applied to the A-coaction on x; in other words, the cokernel of P_k^0 is $s^{|\xi_k|} P(n)_k$, as an A-comodule. This finishes the proof of part (b).

Part (c): This is similar to (b); see also the proof of Lemma 1.2.15(b). By the proof of (a), multiplication by the element $P_k^0 \in A^*$ is a map of A-comodules $P(n)_{k+1} \to P(n)_{k+1}$; therefore, so is multiplication by $(P_k^0)^{p-1}$. The kernel of P_k^0 is $P(n)_k$; the kernel of $(P_k^0)^{p-1}$ is another A-comodule, M. As in part (b), one can identify the cokernels of P_k^0 and $(P_k^0)^{p-1}$ as suspensions of M and $P(n)_k$, respectively, yielding the advertised short exact sequences. As in the proof of Lemma 1.2.15(b), one can identify the Yoneda product $\beta\alpha$ as b_{k0}. □

We need to use the "pth power" functor Φ on the category of A-comodules. When p is odd, we assume that M is an evenly graded comodule; when $p = 2$, M may be arbitrary. ΦM is defined to be

$$(\Phi M)_n = \begin{cases} M_r & n = pr, \\ 0 & n \text{ not divisible by } p, \end{cases}$$

with comodule structure $\psi_{\Phi M} \colon \Phi M \to A \otimes \Phi M$ given by

$$\psi_{\Phi M}(x) = (F \otimes i) \circ \psi_M(i^{-1}x).$$

Here, F is the Frobenius map on A defined by $F(a) = a^p$, ψ_M is the comodule structure map on M, and $i \colon M \to \Phi M$ is the apparent vector space isomorphism (which multiplies degrees by p).

PROPOSITION 4.5.7. *For any prime p, $P(n)_1 \cong \Phi P(n-1)$ as A-comodules.*

This proposition is the main technical result here; its proof requires a careful analysis of the A-comodule structure on $P(n)$. We prove this in Subsection 4.5.1.

We can now prove the main theorem of this section.

PROOF OF THEOREM 4.5.1. Part (a): Corollary 4.5.4 lets us describe $\frac{1}{2}A(n)$ as either $P(n)_{n+1}$ (when $p = 2$) or $A(n)_n$ (otherwise).

Suppose that p is odd. To identify the $Z(d)$-homology groups of $\frac{1}{2}A(n)$, we start with the computation of the $Z(d)$-homology groups of $A(n)_{n+1} = A(n)$ in Theorem 2.3.6(c): when $\xi_t^{p^s}$ is nonzero in $A(n)$, then $(P_t^s)_{**}A(n) = 0$; and when τ_i is nonzero in $A(n)$, then $(Q_i)_{**}A(n) = 0$.

On the comodule $\frac{1}{2}A(n) = A(n) \square_E \mathbf{F}_p$, the element Q_n—the dual of τ_n—acts trivially. Hence $H(\frac{1}{2}A(n), Q_n) \neq 0$. Now assume that $d \neq a_n$. The short exact sequence

$$0 \to \frac{1}{2}A(n) \to A(n) \to s^{a_n}\frac{1}{2}A(n) \to 0$$

gives a long exact sequence in $Z(d)$-homology; since $A(n)$ is $Z(d)_{**}$-acyclic, there is an isomorphism

$$Z(d)_{**}\Sigma^{0,a_n}\frac{1}{2}A(n) \xrightarrow[\cong]{Z(d)_{**}v_n} Z(d)_{**}\Sigma^{-1,0}\frac{1}{2}A(n).$$

Composing this map with itself and using the periodicity in $Z(d)_{**}$, we end up with

$$Z(d)_{i,j}\frac{1}{2}A(n) \cong Z(d)_{i+2,j+2a_n}\frac{1}{2}A(n)$$
$$\cong Z(d)_{i,j+2a_n-2d}\frac{1}{2}A(n)$$
$$\cong Z(d)_{i,j+2\ell(a_n-d)}\frac{1}{2}A(n) \quad \forall \ell \in \mathbf{Z}.$$

Since $\frac{1}{2}A(n)$ is connective and $a_n - d$ is nonzero, we conclude that $Z(d)_{**}\frac{1}{2}A(n) = 0$. (See Proposition 2.2.9 for a similar argument, including a few more details.)

As far as showing that $Z(d)_{**}\frac{1}{2}A(n)$ is nonzero for various d's, if $\xi_t^{p^s} = 0$ in $A(n)$, then a Poincaré series argument shows that at the comodule level, $H(\frac{1}{2}A(n), P_t^s) \neq 0$. Similarly, $H(\frac{1}{2}A(n), Q_i) \neq 0$ when $\tau_i = 0$ in $A(n)$.

The same argument works at the prime 2, after replacing $A(n)$ by $P(n)$.

Part (b): Writing $A(n)_k$ for both the comodule and its injective resolution, we have a sequence of spectra

$$A(n)_{n+1} = A(n), \ A(n)_n, \ A(n)_{n-1}, \ A(n)_{n-2}, \ \ldots, \ A(n)_1,$$
$$A(n)_0 = P(n-1), \ P(n-1)_{n+1}, \ P(n-1)_n, \ \ldots, \ P(n-1)_2,$$
$$P(n-1)_1 = \Phi P(n-2), \ \Phi P(n-2)_n, \ \Phi P(n-1)_{n-1}, \ \ldots, \ \Phi P(n-2)_2,$$
$$\Phi P(n-2)_1 = \Phi^2 P(n-3), \ \ldots,$$
$$\ldots$$
$$\Phi^n P(0), \ \Phi^n P(0)_1 = S^0,$$

together with cofibrations connecting them, of the following forms when p is odd:

$$A(n)_k \to A(n)_{k+1} \to \Sigma^{0,|\tau_k|}A(n)_k \xrightarrow{v_k} \Sigma^{-1,0}A(n)_k,$$
$$\Phi^i P(n-i)_k \to \widetilde{X}_{k+1} \to \Sigma^{1,2p|\xi_k^{p^i}|}\Phi^i P(n-i)_k \xrightarrow{b_{ki}} \Sigma^{-1,0}\Phi^i P(n-i)_k,$$

where the spectrum $\widetilde{X}_{k+1}$ is the cofiber of a nilpotent self-map of $\Phi^i P(n-i)_{k+1}$ (cf. Lemma 1.2.15(b) and Corollary 1.5.2(b)). We should point out that Φ is a functor on A-comodules, not on spectra, so $\Phi^i P(n-i)_k$ means an injective resolution of the comodule $P(n-i)_k$, not some functor Φ^i applied to the spectrum $P(n-i)_k$.

When $p = 2$, we have cofibrations

$$\Phi^i P(n-i)_k \to \Phi^i P(n-i)_{k+1} \to \Sigma^{0,|\xi_k^{2^i}|}\Phi^i P(n-i)_k \xrightarrow{h_{ki}} \Sigma^{-1,0}\Phi^i P(n-i)_k.$$

For any prime p, we can use these cofibrations to imitate the proof of Theorem 2.5.3(a). □

We still owe the proof of Proposition 4.5.7, which we have put into its own subsection.

4.5.1. The proof of Proposition 4.5.7.
Recall from Definition 4.3.3 that we have Hopf algebras

$$P(n) = \mathbf{F}_p[\xi_1, \xi_2, \xi_3, \ldots]/(\xi_1^{p^{n+1}}, \xi_2^{p^n}, \ldots, \xi_n^{p^2}, \xi_{n+1}^p, \xi_{n+2}, \xi_{n+3}, \ldots),$$
$$B(1) = \mathbf{F}_2[\xi_1, \xi_2, \xi_3, \ldots]/(\xi_i^p : i \geq 1).$$

Recall also that $P(n)$ is an A-comodule, and that

$$P(n)_1 = \ker\Big(P(n) \to JA \otimes P(n) \to JB(1) \otimes P(n)\Big).$$

Our goal in this subsection is to show that $P(n)_1$ is isomorphic, as an A-comodule, to $\Phi P(n-1)$, the "Frobenius" operator applied to $P(n-1)$.

By Proposition 4.5.6(a), we know that $P(n)_1$ and $\Phi P(n-1)$ are isomorphic as vector spaces, and indeed as $\Phi P(n-1)$-comodules. As noted after Remark 4.3.4, this does not guarantee that they are isomorphic as A-comodules. Thus we want to show that the A-comodule structure on $P(n)_1$ is Φ applied to Mitchell's and Smith's structure on $P(n-1)$.

We start by making some reductions. We want to show that there is a short exact sequence of A-comodules
$$0 \to \Phi P(n-1) \to P(n) \to P(n)/P(n)_1 \to 0.$$
It is enough to show that there is an A-comodule monomorphism $i\colon \Phi P(n-1) \to P(n)$: since $\Phi P(n-1)$ is nonzero only in degrees divisible by p, then the cokernel of such a map must contain every ξ_k, $1 \le k \le n+1$, and hence there will be an induced surjection $P(n)/\operatorname{cok} i \twoheadrightarrow P(n)/P(n)_1$. Since $P(n)_1$ and $\Phi P(n-1)$ have the same dimension, this map must be an isomorphism.

To construct $i\colon \Phi P(n-1) \to P(n)$, we use Jeff Smith's description of the A-module structure on $P(n)$ as given by Mitchell in [**Mit86**]: for each integer $n \ge 1$, we let Z_n denote the n-dimensional A-comodule with vector space basis
$$(y_1, y_p, y_{p^2}, \ldots, y_{p^{n-1}}),$$
with $|y_{p^i}|$ equal to p^i when $p = 2$, or $2p^i$ when p is odd, and with comodule structure given by
$$y_{p^i} \mapsto \sum_{j=0}^{i} \xi_{i-j}^{p^j} \otimes y_{p^j}.$$
At the prime 2, for example, if we let $Z = \operatorname{Span}(y_1, y_2, y_4, \ldots)$ be the quotient comodule of $H_*(\mathbf{R}P^\infty; \mathbf{F}_2)$ "generated" by the class in degree 1, then Z_n is an n-dimensional subcomodule of Z. For any non-negative integer k, we write Σ_k for the symmetric group on k letters, and we let e_n be a certain idempotent (defined below) in the group algebra $\mathbf{F}_p[\Sigma_{(p-1)\binom{n}{2}}]$.

Note that Z_{n+2} is an A-comodule, and the A-coaction commutes with the symmetric group action on $Z_{n+2}^{\otimes(p-1)\binom{n+2}{2}}$, so $Z_{n+2}^{\otimes(p-1)\binom{n+2}{2}} e_{n+2}$ inherits an A-comodule structure. By restriction, it has a $P(n)$-comodule structure, as well.

We have the following.

THEOREM 4.5.8 (Smith, [**Mit86**]). $Z_{n+2}^{\otimes(p-1)\binom{n+2}{2}} e_{n+2}$ *is cofree of rank one as a $P(n)$-comodule. It is self-dual as an A-comodule.* $Z_n^{\otimes(p-1)\binom{n}{2}} e_n$ *has bottom class in degree* $2(\frac{p^n-1}{p-1} - n)$ *when p is odd, and in degree* $2^n - 1 - n$ *when $p = 2$.*

(The self-duality and the identification of the degree of the bottom class come from Mitchell's construction, [**Mit85**, Theorem 3.1], combined with Mitchell's isomorphism between the two A-comodule structures, [**Mit86**, Theorem 4.2].)

Since the A-comodule structure on $P(n)$ is self-dual, we have the following.

COROLLARY 4.5.9. *Up to suspension, there is an isomorphism of A-comodules*:
$$Z_{n+2}^{\otimes(p-1)\binom{n+2}{2}} e_{n+2} \cong (Z_{n+2}^*)^{\otimes(p-1)\binom{n+2}{2}} e_{n+2}.$$
Hence $(Z_{n+2}^*)^{\otimes(p-1)\binom{n+2}{2}} e_{n+2}$ is cofree of rank one as a $P(n)$-comodule.

We note that ΦZ_{n+1}^* is a subcomodule of Z_{n+2}^*. In order to construct the map $i\colon \Phi P(n-1) \hookrightarrow P(n)$, we prove in Proposition 4.5.11 that there is an isomorphism of A-comodules (neglecting suspensions)
$$Z_{n+1}^{\otimes(p-1)\binom{n+1}{2}} e_{n+1} \cong Z_{n+1}^{\otimes(p-1)\binom{n+2}{2}} e_{n+2}.$$
Hence the inclusion $\Phi Z_{n+1}^* \hookrightarrow Z_{n+2}^*$ induces an injective map
$$(\Phi Z_{n+1}^*)^{\otimes(p-1)\binom{n+2}{2}} e_{n+2} \hookrightarrow (Z_{n+2}^*)^{\otimes(p-1)\binom{n+2}{2}} e_{n+2},$$

4.5. $\frac{1}{2}A(n)$ GENERATES THE EXPECTED THICK SUBCATEGORY

which is the map i that we want. (We remark that one can check that the bottom class in the domain is in the same degree as the bottom class in codomain, but one can see more from general principles: the map is $P(n)$-linear, and any nonzero $P(n)$-comodule map into $P(n)$ must have the bottom class in the image, by cofreeness. Since the map is injective, we conclude that it must send $1 \in \Phi P(n-1)$ to $1 \in P(n)$.)

Now we discuss the details of the construction, starting with the description of the idempotent $e_n \in \mathbf{F}_p[\Sigma_{(p-1)\binom{n}{2}}]$. James and Kerber [**JK81**] have written one of the standard references for representations of the symmetric group; in [**Mit86**], Mitchell provides most of the results needed to study Smith's construction.

In general, idempotents for the symmetric group Σ_k come from a partition of k; in our case, we consider the partition
$$\lambda_n = ((n-1)(p-1), (n-2)(p-1), \ldots, 2(p-1), (p-1))$$
of $(p-1)\binom{n}{2}$. Associated to the partition λ_n is a *Young diagram of shape* λ_n, as described in [**JK81**, Section 4]. Given a vector space V together with a basis for V, we can view a standard basis element of $V^{\otimes (p-1)\binom{n}{2}}$ as a *Young tableau*—a Young diagram of shape λ_n in which each box is filled in with a basis element of V. For example, if $p = 2$, $n = 3$, and $\{u, v, w\}$ is a basis for V, then the tableau $\begin{array}{|c|c|}\hline u & v \\\hline w \\\cline{1-1}\end{array}$ represents the element $u \otimes v \otimes w$ in $V^{\otimes 3}$. An arbitrary element of $V^{\otimes (p-1)\binom{n}{2}}$ is a sum of such tableaux.

The idempotent element $e_n \in \mathbf{F}_p[\Sigma_{(p-1)\binom{n}{2}}]$ is defined as follows: we let $R = R_n$ be the row-stabilizer subgroup of $\Sigma_{(p-1)\binom{n}{2}}$ associated to the Young diagram of shape λ_n—in other words, the subgroup of permutations of $(p-1)\binom{n}{2}$ that send each row of λ to itself—and we let $\overline{R}$ be the sum of the elements in R. Similarly, we let $C = C_n$ be the column stabilizer subgroup and $\widetilde{C}$ the signed sum of the elements in C. We let $f_n = \overline{R}\widetilde{C}$. One can show (e.g., by applying [**JK81**, Theorem 3.1.10] and noting that the hook lengths of the diagram are all prime to p) that $f_n^2 = \alpha f_n$ for some nonzero scalar $\alpha \in \mathbf{F}_p$, so $e_n = \frac{1}{\alpha} f_n$ is an idempotent in $\mathbf{F}_p[\Sigma_{(p-1)\binom{n}{2}}]$.

Results from the theory of representations of the symmetric groups give the following (cf. [**Mit86**, Lemma 1.2]):

LEMMA 4.5.10. *Let V be a vector space over a field k of characteristic p. Then*
$$\dim_k(V^{\otimes (p-1)\binom{n}{2}} e_n) = \begin{cases} 0 & \text{if } \dim_k V < n-1, \\ p^{\binom{n-1}{2}} & \text{if } \dim_k V = n-1, \\ p^{\binom{n}{2}} & \text{if } \dim_k V = n. \end{cases}$$

SKETCH OF PROOF. Let $W_n = V^{\otimes (p-1)\binom{n}{2}} e_n$. A formula for the dimension of W_n in terms of hook lengths of the partition λ_n is given in [**JK81**, Theorem 5.2.14]. It is clear that if $\dim V < n - 1$, then $\dim W_n = 0$; see [**JK81**, Theorem 5.2.15], for instance. If $\dim V = n - 1$ or $\dim V = n$, then a reasonably straightforward computation using this formula gives the advertised dimension for W_n. (When $\dim V > n$, the dimension formula is more complicated.) □

Thus we have
$$\dim Z_{n-1}^{\otimes (p-1)\binom{n-1}{2}} e_{n-1} = \dim Z_{n-1}^{\otimes (p-1)\binom{n}{2}} e_n = \dim P(n-3).$$

As remarked after Corollary 4.5.9, Proposition 4.5.7 follows if we know not just that these vector spaces have the same dimension, but that they are isomorphic as A-comodules. Reindexing, we have to prove this result.

PROPOSITION 4.5.11. *As A-comodules, $Z_n^{\otimes(p-1)\binom{n}{2}}e_n \cong Z_n^{\otimes(p-1)\binom{n+1}{2}}e_{n+1}$, up to suspension.*

COROLLARY 4.5.12. (a) *There is an A-linear injection $P(n-1) \hookrightarrow P(n)$; dually, there is an A-linear surjection $P(n) \twoheadrightarrow s^{c_n}P(n-1)$, where c_n is chosen so that the top class of $P(n)$ gets sent to the top class of $P(n-1)$.*
(b) *There is an A-linear injection $\Phi P(n-1) \hookrightarrow P(n)$; dually, there is an A-linear surjection $P(n) \twoheadrightarrow s^{b_n}\Phi P(n-1)$.*

Part (a) of this corollary is originally due to Mitchell [**Mit85**, Corollary 3.13].

PROOF OF PROPOSITION 4.5.11. Define a map $\varphi \colon Z_n^{\otimes(p-1)\binom{n}{2}} \to Z_n^{\otimes(p-1)\binom{n+1}{2}}$ by

$$\varphi((a_{1,1} \otimes \cdots \otimes a_{1,(p-1)(n-1)}) \otimes \cdots \otimes (a_{n-1,1} \otimes \cdots \otimes a_{n-1,p-1})) =$$
$$\sum_{\sigma \in \Sigma_n} \operatorname{sgn}(\sigma)(y_{p^{\sigma(0)}}^{\otimes p-1} \otimes a_{1,1} \otimes \cdots \otimes a_{1,(p-1)(n-1)})$$
$$\otimes \cdots \otimes (y_{p^{\sigma(n-1)}}^{\otimes p-1} \otimes a_{n-1,1} \otimes \cdots \otimes a_{n-1,p-1}).$$

(Here Σ_n is permuting the set $\{0, 1, \ldots, n-1\}$.) In other words, φ takes a tableau T of shape λ_n to the signed sum of $n!$ tableaux of shape λ_{n+1}, one for each $\sigma \in \Sigma_n$. The new tableau corresponding to σ is obtained by adding $p-1$ columns of height n to the left of T, by putting $y_{p^{\sigma(i)}}$ in each new box in the ith row, and by multiplying that new tableau by $\operatorname{sgn}(\sigma)$. For example, when $p = 2$ and $n = 3$, φ is given by

$$\begin{array}{|c|c|}\hline a & b \\\hline c & \\\hline\end{array} \longmapsto \begin{array}{|c|c|c|}\hline 1 & a & b \\\hline 2 & c & \\\hline 4 & & \\\hline\end{array} + \begin{array}{|c|c|c|}\hline 1 & a & b \\\hline 4 & c & \\\hline 2 & & \\\hline\end{array} + \begin{array}{|c|c|c|}\hline 2 & a & b \\\hline 1 & c & \\\hline 4 & & \\\hline\end{array} + \begin{array}{|c|c|c|}\hline 2 & a & b \\\hline 4 & c & \\\hline 1 & & \\\hline\end{array} + \begin{array}{|c|c|c|}\hline 4 & a & b \\\hline 1 & c & \\\hline 2 & & \\\hline\end{array} + \begin{array}{|c|c|c|}\hline 4 & a & b \\\hline 2 & c & \\\hline 1 & & \\\hline\end{array}.$$

Here we are writing k for the basis element $y_k \in Z_n$. φ is clearly $\mathbf{F}_p$-linear and monomorphic. φ is also a map of A-comodules: to see this, let

$$b = \sum_{\sigma \in \Sigma_n} \operatorname{sgn}(\sigma) y_{p^{\sigma(0)}}^{\otimes p-1} \otimes \cdots \otimes y_{p^{\sigma(n-1)}}^{\otimes p-1}.$$

By commutativity of A, φ is A-linear if and only if the map

$$\psi \colon Z_n^{\otimes(p-1)\binom{n}{2}} \longrightarrow Z_n^{\otimes(p-1)\binom{n+1}{2}}$$
$$T \longmapsto b \otimes T$$

is A-linear. Moreover, ψ is A-linear if and only if b is primitive in $Z_n^{\otimes(p-1)n}$. By commutativity of A, though, it is: one can easily see that under the A-coaction, b goes to $1 \otimes b$.

Hence we have a monomorphism of A-comodules,

$$\varphi \colon Z_n^{\otimes(p-1)\binom{n}{2}} \to Z_n^{\otimes(p-1)\binom{n+1}{2}}.$$

The idempotents e_n and e_{n+1} split each into summands as A-comodules, so we have an induced map

$$\overline{\varphi} \colon Z_n^{\otimes(p-1)\binom{n}{2}}e_n \to Z_n^{\otimes(p-1)\binom{n+1}{2}}e_{n+1}.$$

Since these have the same vector space dimension and since φ is monomorphic, it suffices to check that $\overline{\varphi}$ is nonzero, so it suffices to find one element in the image of e_n which gets sent to a nonzero element in the image of e_{n+1}.

This last step is essentially done by Mitchell in [**Mit86**, Section 2]; at least, Mitchell identifies elements in the image of e_n. One can check by induction that if T is the tableau of shape λ_n with rows of the form
$$y_{p^i}^{\otimes p-1} \otimes y_{p^{i-1}}^{\otimes p-1} \otimes \cdots \otimes y_p^{\otimes p-1} \otimes y_1^{\otimes p-1},$$
then $T \cdot \widetilde{C}_n$ is a nonzero element of $Z_n^{\otimes (p-1)\binom{n}{2}} e_n$, and $\overline{\varphi}(T \cdot \widetilde{C}_n)$ is a nonzero element of $Z_n^{\otimes (p-1)\binom{n+1}{2}} e_{n+1}$. □

(Indeed, the bottom class of $Z_n^{\otimes (p-1)\binom{n}{2}} e_n$ is the tableau $T \cdot \widetilde{C}_n$, and hence similarly for $Z_n^{\otimes (p-1)\binom{n+1}{2}} e_{n+1}$ and $\overline{\varphi}(T \cdot \widetilde{C}_n)$.)

4.6. Some computations and applications

In this section, we compute $(Q_n)_{**}(Q_n)$ and give several applications: we show that $v_0^{-1} \pi_{**} X \cong (Q_0)_{**} X$ for every connective X, we reproduce Eisen's calculation of a particular localized Ext group, and we examine Mahowald's conjectured calculation of the v_1-inverted Ext of the Moore spectrum.

4.6.1. Computation of $(Q_n)_{**}(Q_n)$. In this subsection, we compute the Hopf algebra $(Q_n)_{**}(Q_n)$ of $(Q_n)_{**}$-operations for all n, and we use this to compute $v_0^{-1} \pi_{**} X$ when X is connective.

We start by computing $(q_n)_{**}(q_n)$. Margolis computed this when $p = 2$; see [**Mar83**, Corollary 19.26] for his version. We extend his work to all primes in the following proposition.

For an A-comodule M, we write $H(M, Q_n)$ for the conventional Q_n-homology of M,
$$H(M, Q_n) = \frac{\ker(M \xrightarrow{Q_n} M)}{\operatorname{im}(M \xrightarrow{Q_n} M)},$$
as discussed at the beginning of Section 2.2. Recall from Notation 4.1.1 that the Hopf algebra E is defined by
$$E = \begin{cases} \mathbf{F}_2[\xi_{n+1}]/(\xi_{n+1}^2), & \text{when } p = 2, \\ E[\tau_n], & \text{when } p \text{ is odd.} \end{cases}$$

PROPOSITION 4.6.1. *Fix $n \geq 0$.*
- (a) $(q_n)_{**}(q_n) = HE_{**}HE \cong \operatorname{Ext}_E^{**}(\mathbf{F}_p, A \square_E \mathbf{F}_p)$.
- (b) *As an E-comodule, $A \square_E \mathbf{F}_p$ is a direct sum of trivial comodules and cofree comodules. The trivial summands are indexed by $H(A \square_E \mathbf{F}_p, Q_n)$, and the cofree summands are indexed by $\operatorname{im}(Q_n \colon A \square_E \mathbf{F}_p \to A \square_E \mathbf{F}_p)$.*
- (c) *Hence $(q_n)_{**}(q_n)$ is a direct sum of copies of $\mathbf{F}_p[v_n]$ and $\mathbf{F}_p$. The $\mathbf{F}_p[v_n]$ summands are indexed by $H(A \square_E \mathbf{F}_p, Q_n)$, and the $\mathbf{F}_p$ summands are indexed by $\operatorname{im} Q_n$.*
- (d) *Let C_n be the quotient Hopf algebra of A defined by $C_n = A/(\xi_k^{p^n} : k \geq 1)$.*

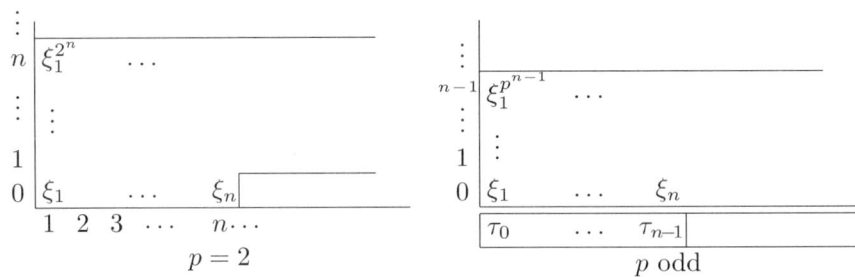

FIGURE 4.6.A. "Profile function" for $H(A \square_E \mathbf{F}_p, Q_n)$. Although $H(A \square_E \mathbf{F}_p, Q_n)$ is not a quotient of A, it still makes sense to show which $\xi_i^{p^j}$'s and which τ_i's are nonzero in it.

(i) When $p = 2$, $H(A \square_E \mathbf{F}_2, Q_n)$ is isomorphic to the sub-Hopf algebra $C_{n+1} \square_{B(n+1)} \mathbf{F}_2$ of C_{n+1}, where $B(n+1)$ is the quotient Hopf algebra
$$B(n+1) = \mathbf{F}_2[\xi_{n+1}, \xi_{n+2}, \ldots]/(\xi_k^2 : k \geq n+1).$$

(ii) When p is odd, $H(A \square_E \mathbf{F}_p, Q_n)$ is isomorphic to the sub-Hopf algebra $C_n \square_{\Lambda(n)} \mathbf{F}_p$ of C_n, where $\Lambda(n)$ is the quotient Hopf algebra
$$\Lambda(n) = \Lambda[\tau_n, \tau_{n+1}, \ldots].$$

In both cases, $\operatorname{im} Q_n$ is nonzero. For example, $\xi_1^{p^{n+1}}$ is in $\operatorname{im} Q_n$ for all primes p.

See Figure 4.6.A for the profile function of $H(A \square_E \mathbf{F}_p, Q_n)$.

PROOF. Part (a): This is an application of Shapiro's Lemma—see Example 1.2.4, and also Proposition 2.2.3.

Part (b): This is a basic fact about comodules over an exterior coalgebra, applied to the particular comodule $A \square_E \mathbf{F}_p$, using the fact that the element Q_n of A^* is dual to either ξ_{n+1} ($p = 2$) or τ_n (p odd). See Corollary 0.5.5.

Part (c): This is a basic fact about Ext over an exterior coalgebra. See Proposition 0.5.10.

Part (d): This is the interesting part. Since Q_n is primitive in A^*, if we know how it acts on the algebra generators of $A \square_E \mathbf{F}_p$, then we can understand the action on the whole algebra. Indeed, if we know how Q_n acts on the generators of A itself, then we can understand how it acts on $A \square_E \mathbf{F}_p$.

Let $p = 2$. Since Q_n is dual to ξ_{n+1}, then multiplication by Q_n does the following on the generators of A:
$$\begin{aligned} \xi_k &\mapsto 0, & \text{if } k < n+1, \\ \xi_{n+1} &\mapsto 1, \\ \xi_k &\mapsto \xi_{k-n-1}^{2^{n+1}}, & \text{if } k > n+1. \end{aligned}$$

Therefore, given any monomial in A, if it contains an odd power of any ξ_k with $k \geq n+1$, then it is not in $A \square_E \mathbf{F}_p$. The cotensor product is spanned by all other monomials. The description of the action of Q_n, together with the primitivity of Q_n, allows us to identify $H(A \square_E \mathbf{F}_p, Q_n)$.

Similarly, when p is odd, multiplication by Q_n does this:

$$\begin{aligned} \tau_k &\mapsto 0, &&\text{if } k < n, \\ \tau_n &\mapsto 1, \\ \tau_k &\mapsto \xi_{k-n}^{p^n}, &&\text{if } k > n, \\ \xi_k &\mapsto 0, &&\text{for all } k. \end{aligned}$$

$\square$

See Example 4.6.4 for some computations for small values of n.

COROLLARY 4.6.2. q_n is not a flat ring spectrum.

PROOF. The trivial module $\mathbf{F}_p$ is not flat over the polynomial ring $(q_n)_{**} = \mathbf{F}_p[v_n]$, so one only has to check that $\operatorname{im} Q_n$ is not zero. This is part (d) of the proposition. $\square$

We can use Proposition 4.6.1 to compute $(Q_n)_{**}(Q_n)$.

PROPOSITION 4.6.3. *Fix* $n \geq 0$. *Then*

$$\begin{aligned} (Q_n)_{**}(Q_n) &\cong v_n^{-1}(q_n)_{**}(q_n) \\ &\cong v_n^{-1}\operatorname{Ext}_E^{**}(\mathbf{F}_p, A \square_E \mathbf{F}_p) \\ &\cong \mathbf{F}_p[v_n^{\pm 1}] \otimes H(A \square_E \mathbf{F}_p, Q_n). \end{aligned}$$

PROOF. We only need to verify that $(Q_n)_{**}(Q_n) \cong v_n^{-1}(q_n)_{**}(q_n)$, but this is clear, once we note that $Q_n = v_n^{-1}q_n$. The rest follows from Proposition 4.6.1. $\square$

EXAMPLE 4.6.4. Here are some computations of the Q_n-homology of $A \square_E \mathbf{F}_p$, along with the corresponding coalgebras of operations.

(a) Consider Q_0. For all primes p, $H(A \square_E \mathbf{F}_p, Q_0) = \mathbf{F}_p$, so $(Q_0)_{**}(Q_0) = \mathbf{F}_p[v_0^{\pm 1}]$. This is a graded field, so $\operatorname{Ext}^s_{(Q_0)_{**}(Q_0)}(-,-) = 0$ when $s > 0$.
(b) Consider Q_1. If $p = 2$, then $C_2 = A/(\xi_k^4)$ and $B(2) = \mathbf{F}_2[\xi_2, \xi_3, \ldots]/(\xi_k^2)$, so

$$H(A \square_E \mathbf{F}_2, Q_1) \cong \mathbf{F}_2[\xi_1, \xi_2^2, \xi_3^2, \ldots, \xi_i^2, \ldots]/(\xi_k^4).$$

When p is odd,

$$H(A \square_E \mathbf{F}_p, Q_1) \cong E[\tau_0] \otimes \mathbf{F}_p[\xi_1, \xi_2, \ldots]/(\xi_k^p).$$

In either case, one obtains $(Q_1)_{**}(Q_1)$ from this by tensoring with $\mathbf{F}_p[v_1^{\pm 1}]$. It is often easy to compute Ext over these coalgebras. For example, if M is the homology of the mod p Moore spectrum, $M = \mathbf{F}_2[\xi_1]/(\xi_1^2)$ or $M = E[\tau_0]$, then $\operatorname{Ext}^{**}_{(Q_1)_{**}(Q_1)}((Q_1)_{**}, (Q_1)_{**}M)$ is given by

$$\begin{aligned} &\mathbf{F}_2[v_1^{\pm 1}] \otimes \mathbf{F}_2[h_{11}, h_{21}, h_{31}, \ldots], &&\text{if } p = 2, \\ &\mathbf{F}_p[v_1^{\pm 1}] \otimes \Lambda[h_{10}, h_{20}, h_{30}, \ldots] \otimes \mathbf{F}_p[b_{10}, b_{20}, b_{30}, \ldots], &&\text{if } p \text{ is odd.} \end{aligned}$$

(c) For all n, one can give a similar description of $(Q_n)_{**}(Q_n)$; however, the larger n gets, the more difficult it gets to compute its cohomology.

Combining the computation of $(Q_0)_{**}(Q_0)$ with the vanishing plane condition (4.2.7) yields the following.

COROLLARY 4.6.5. *For every connective spectrum X, the Q_0-based Adams spectral sequence abutting to $v_0^{-1}\pi_{**}X$ converges. Furthermore, the spectral sequence collapses at E_2, and $v_0^{-1}\pi_{**}X = (Q_0)_{**}X$.*

PROOF. We claim that the vanishing plane condition (4.2.7), with $\varepsilon = 1$, is satisfied at the E_1-term of the spectral sequence. Note that the slope of Q_0 is 1, so the condition is that for some number b,
$$E_1^{s,t,u} = 0 \text{ when } 2s \geq t + b.$$
We give a construction of the Adams spectral sequence in Section 1.4; from there, we recall that
$$E_1^{s,t,u} = \pi_{s+u,t+u}(q_0 \wedge \overline{q_0}^{\wedge s} \wedge X).$$
Since $S^0 \to q_0$ is an epimorphism on π_{**}, and an isomorphism on π_{ij} when $j - i \leq 0$ and also when $i \leq 0$, then $\pi_{ij}(\overline{q_0})$ is zero when $j - i \leq 0$ and when $i \leq 0$. By the Hurewicz theorem 1.3.5 and the Künneth theorem 1.2.11 for mod p homology $H\mathbf{F}_p$, we find that if X is (i_0, j_0)-connective (Definition 1.3.4), then
$$\pi_{s+u,t+u}(q_0 \wedge \overline{q_0}^{\wedge s} \wedge X) = 0 \text{ when } (t+u) - (s+u) < j_0 + s.$$
In other words, $E_1^{s,t,u} = 0$ when $2s > t - j_0$: the vanishing plane condition holds at E_1 with $\varepsilon = 1$ and $b = -j_0$. So the spectral sequence converges.

We know the E_2-term of the Q_0-based spectral sequence by Theorem 4.2.3; using our computation in Example 4.6.4 that $(Q_0)_{**}(Q_0) \cong (Q_0)_{**} \cong \mathbf{F}_p[v_0^{\pm 1}]$, we have
$$\begin{aligned} E_2^{s,t,u} &\cong \operatorname{Ext}_{\mathbf{F}_p[v_0^{\pm 1}]}^{s,t,u}(\mathbf{F}_p[v_0^{\pm 1}], (Q_0)_{**}X) \\ &\cong \operatorname{Ext}_{\mathbf{F}_p}^{s,t,u}(\mathbf{F}_p, (Q_0)_{**}X) \\ &\cong \begin{cases} (Q_0)_{**}X, & \text{when } s = 0, \\ 0, & \text{when } s \neq 0. \end{cases} \end{aligned}$$
Since the spectral sequence lies completely in the plane $s = 0$, there is no room for any differentials. □

REMARK 4.6.6. If one works over a quotient A' of A in which ξ_{n+1} (respectively, τ_n) is nonzero, one can make similar computations and draw similar conclusions: $(q_n)_{**}(q_n)$ is computed via $H(A' \square_E \mathbf{F}_p, Q_n)$, the action of the element Q_n on $A' \square_E \mathbf{F}_p$ can be computed by the same formulas, and $(Q_n)_{**}(Q_n) = v_n^{-1}(q_n)_{**}(q_n)$.

As a consequence, if no ξ_i in A' is nilpotent, then the computation of the homology group $H(A' \square_E \mathbf{F}_p, Q_n)$ is much like the computations above. If there is torsion, though, things can be more complicated. For example, when $p = 2$ and $n = 0$, then in $A \square_E \mathbf{F}_2$, we have $Q_0(\xi_1^2 \xi_2) = \xi_1^4$. On the other hand, $\xi_1^4 = 0$ in $A(1)$, so in $A(1) \square_E \mathbf{F}_2$, we have $Q_0(\xi_1^2 \xi_2) = 0$. Hence over $A(1)$, $(Q_0)_{**}(Q_0) \cong \mathbf{F}_2[v_1^{\pm 1}] \otimes E[x_5]$, where $|x_5| = 5$. Hence $v_0^{-1} \operatorname{Ext}_{A(1)}^{**}(\mathbf{F}_2, \mathbf{F}_2) \cong \mathbf{F}_2[v_0^{\pm 1}, y_4]$, where y_4 is "in the 4-stem": y_4 has bidegree $(1, 5)$.

4.6.2. Eisen's calculation. Let $Y(n)$ be the quotient Hopf algebra of A defined by
$$Y(n) = \begin{cases} \mathbf{F}_2[\xi_{n+1}, \xi_{n+2}, \xi_{n+3}, \dots], & \text{if } p = 2, \\ \mathbf{F}_p[\xi_{n+1}, \xi_{n+2}, \xi_{n+3}, \dots] \otimes \Lambda[\tau_n, \tau_{n+1}, \dots], & \text{if } p \text{ is odd.} \end{cases}$$
By Proposition 1.2.8, $HY(n)_{**} \cong \operatorname{Ext}_{Y(n)}^{**}(\mathbf{F}_p, \mathbf{F}_p)$. Since ξ_{n+1} (respectively, τ_n) is primitive in $Y(n)$, then v_n is a class in $HY(n)_{1,a_n}$. This class is non-nilpotent because it is detected by the restriction map associated to the quotient map $Y(n) \to E$.

Eisen proved the following in [**Eis87**].

THEOREM 4.6.7. *Fix $n \geq 0$.*

(a) *When $p = 2$,*
$$v_n^{-1} \operatorname{Ext}_{Y(n)}^{**}(\mathbf{F}_2, \mathbf{F}_2) \cong \mathbf{F}_2[v_n^{\pm 1}] \otimes \mathbf{F}_2[h_{t0} : n+1 < t < 2n+2]$$
$$\otimes \mathbf{F}_2[h_{ts} : 0 < s < n+1, \ t \geq n+1].$$

(b) *When p is odd,*
$$v_n^{-1} \operatorname{Ext}_{Y(n)}^{**}(\mathbf{F}_p, \mathbf{F}_p) \cong \mathbf{F}_p[v_n^{\pm 1}] \otimes \mathbf{F}_p[v_i : n+1 \leq i \leq 2n]$$
$$\otimes \Lambda[h_{ts} : 0 \leq s \leq n-1, \ t \geq n+1]$$
$$\otimes \mathbf{F}_p[b_{ts} : 0 \leq s \leq n-1, \ t \geq n+1].$$

Remarks 4.1.3, 4.2.8, 4.3.2, and 4.6.6 come into play now: we work over the quotient Hopf algebra $Y(n)$ via $HY(n)$. Since v_n is an element in $HY(n)_{**}$, then $S^0 \wedge HY(n)$ has a v_n-map, so by Remarks 4.1.3 and 4.2.8, we can consider the Q_n-based Adams spectral sequence abutting to $v_n^{-1} HY(n)_{**}$.

Remark 4.6.6 tells us how to compute the coalgebra of operations, and hence the E_2-term.

PROPOSITION 4.6.8. *Consider the Q_n-based Adams spectral sequence abutting to $v_n^{-1} HY(n)_{**}$.*

(a) *Suppose that $p = 2$. Let*
$$B = \mathbf{F}_2[\xi_{n+1}, \ldots, \xi_{2n+1}, \xi_{2n+2}^2, \xi_{2n+3}^2, \ldots]/(\xi_k^{2^{n+1}} : k \geq n+1).$$
Then the E_2-term is
$$E_2 \cong \mathbf{F}_2[v_n^{\pm 1}] \otimes \operatorname{Ext}_B^{**}(\mathbf{F}_2, \mathbf{F}_2)$$
$$\cong \mathbf{F}_2[v_n^{\pm 1}] \otimes \mathbf{F}_2[h_{t0} : n+1 < t < 2n+2]$$
$$\otimes \mathbf{F}_2[h_{ts} : 0 < s \leq n, \ t \geq n+1].$$

(b) *Suppose that p is odd. Let*
$$C = \Lambda[\tau_n, \ldots, \tau_{2n}] \otimes \mathbf{F}_p[\xi_{n+1}, \xi_{n+2}, \ldots]/(\xi_k^{p^n} : k \geq n+1).$$
Then the E_2-term is
$$E_2 \cong \mathbf{F}_p[v_n^{\pm 1}] \otimes \operatorname{Ext}_C^{**}(\mathbf{F}_p, \mathbf{F}_p)$$
$$\cong \mathbf{F}_p[v_n^{\pm 1}] \otimes \mathbf{F}_p[v_i : n+1 \leq i \leq 2n]$$
$$\otimes \Lambda[h_{ts} : 0 \leq s \leq n-1, \ t \geq n+1]$$
$$\otimes \mathbf{F}_p[b_{ts} : 0 \leq s \leq n-1, \ t \geq n+1].$$

Remark 4.3.2 tells us about convergence. Since Q_n is the first differential in $Y(n)$, then for every finite spectrum X, $HY(n) \wedge X$ is $Z(d)_{**}$-acyclic for all $d < a_n$; hence the spectral sequence converges for every finite spectrum. (Alternatively, one can check that, as in the proof of Corollary 4.6.5, the vanishing plane condition holds at the E_2-term of the spectral sequence for the sphere.)

COROLLARY 4.6.9. *For any finite spectrum X, the Q_n-based Adams spectral sequence abutting to $v_n^{-1} HY(n)_{**} X$ converges.*

Eisen's theorem 4.6.7 follows from the next lemma.

LEMMA 4.6.10. *The spectral sequence converging to $v_n^{-1} HY(n)_{**}$ collapses at E_2.*

SKETCH OF PROOF. Let $E(n)$ be the quotient Hopf algebra of $Y(n)$ defined by
$$E(n) = \begin{cases} Y(n)/(\xi_k^{2^{n+1}} : k \geq n+1), & \text{if } p = 2, \\ Y(n)/(\xi_k^{p^n} : k \geq n+1), & \text{if } p \text{ is odd.} \end{cases}$$
Just as for $Y(n)$, one can show that the Q_n-based Adams spectral sequence abutting to $v_n^{-1} HE(n)_{**}$ converges. One can also compute $HE(n)_{**}$ directly, by observing that $E(n)$ is a tensor product of Hopf algebras of the forms $\mathbf{F}_p[\xi_k]/(\xi_k^{p^r})$ with ξ_k primitive, and $E[\tau_k]$ with τ_k primitive. Hence:
$$HE(n)_{**} = \begin{cases} \mathbf{F}_2[h_{ts} : 0 \leq s \leq n-1, t \geq n+1], & \text{if } p = 2, \\ \mathbf{F}_p[v_i : i \geq n] \otimes \Lambda[h_{ts} : 0 \leq s \leq n, t \geq n+1] \\ \qquad \otimes \mathbf{F}_p[b_{ts} : 0 \leq s \leq n, t \geq n+1], & \text{if } p \text{ is odd.} \end{cases}$$
Of course, one can compute $v_n^{-1} HE(n)_{**}$ immediately from this.

One can conclude three things: first, the restriction map $HY(n) \to HE(n)$ induces an inclusion at the E_2-terms of the Q_n-based Adams spectral sequence:
$$E_2(HY(n)) \hookrightarrow E_2(HE(n)).$$
Second, the spectral sequence for $E(n)$ collapses at E_2. Thus, third, the same must be true for the $Y(n)$ spectral sequence, completing the proof of the lemma. □

4.6.3. The v_1-inverted Ext of the mod 2 Moore spectrum. Let $p = 2$, let X be the mod 2 Moore spectrum, and let $M = A(0)$ be its homology. In this subsection we discuss Mahowald's unverified computation in [**Mah70**] of $v_1^{-1} \operatorname{Ext}_A^{**}(\mathbf{F}_2, M)$; no proof of this result has been published. Before stating the result, we recall some computations: $\pi_{**}(HA(1) \wedge M)$ is a series of "lightning flashes":
$$\pi_{**}(HA(1) \wedge M) \cong \operatorname{Ext}_A^{**}(\mathbf{F}_2, H_*(bo \wedge X))$$
$$\cong \operatorname{Ext}_{A(1)}^{**}(\mathbf{F}_2, M)$$
$$\cong \mathbf{F}_2[v_1^4] \otimes \mathbf{F}_2[h_{11}, v_1]/(h_{11}^3, v_1^2).$$
Also, if E' is the quotient Hopf algebra $E' = A/(\xi_1^2, \xi_2^2, \xi_3, \xi_4, \dots)$, then $\pi_{**}(HE' \wedge M)$ is a v_1-tower:
$$\pi_{**}(HE' \wedge M) \cong \operatorname{Ext}_A^{**}(\mathbf{F}_2, H_*(bu \wedge X))$$
$$\cong \operatorname{Ext}_{E'}^{**}(\mathbf{F}_2, M)$$
$$\cong \mathbf{F}_2[v_1].$$
See Figure 4.6.B for a few pictures.

To present Mahowald's answer, we also need a definition. Let P be the polynomial algebra $P = \mathbf{F}_2[x_1, x_2, x_3, \dots]$, which is bigraded by setting $|x_i| = (2, 2^{i+2}+1)$. Define a derivation d on P by $d(x_i) = x_1 x_{i-1}^2$. Let $H(d)$ denote the homology with respect to d, and let $B(d)$ denote the image of d.

CONJECTURE 4.6.11 ([**Mah70**]). Let $p = 2$, and let $M = \mathbf{F}_2[\xi_1]/(\xi_1^2)$ be the homology of the mod 2 Moore spectrum. Let $E' = \mathbf{F}_2[\xi_1, \xi_2]/(\xi_1^2, \xi_2^2)$. Then $v_1^{-1} \pi_{**} M$ is a direct sum of copies of $v_1^{-1} \pi_{**}(HA(1) \wedge M)$ and $v_1^{-1} \pi_{**}(HE' \wedge M)$. More

4.6. SOME COMPUTATIONS AND APPLICATIONS

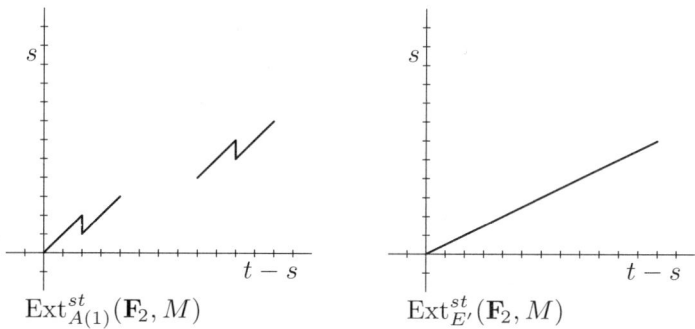

FIGURE 4.6.B. Lightning flashes and v_1-towers. Pictures of $\operatorname{Ext}_{A(1)}^{**}(\mathbf{F}_2, M)$ and $\operatorname{Ext}_{E'}^{**}(\mathbf{F}_2, M)$, where M is the homology of the mod 2 Moore spectrum, $A(1)$ is the usual quotient of A, and $E' = A/(\xi_1^2, \xi_2^2, \xi_3, \xi_4, \dots)$.

precisely, we have:

$$v_1^{-1}\pi_{**}M \cong \bigoplus_{\alpha \in H(d)} \Sigma^{|\alpha|} v_1^{-1} \pi_{**}(HA(1) \wedge M)$$
$$\oplus \bigoplus_{\beta \in B(d)} \Sigma^{|\beta|} v_1^{-1} \pi_{**}(HE' \wedge M).$$

One can try to compute this using the Q_1-based Adams spectral sequence; by a change-of-rings computation, the E_2-term for the object M is

$$E_2 \cong \mathbf{F}_2[v_1^{\pm 1}] \otimes \mathbf{F}_2[h_{11}, h_{21}, h_{31}, \dots].$$

In this case, v_1^4 is a permanent cycle, and the spectral sequence is one of $\mathbf{F}_2[v_1^{\pm 4}]$-modules.

We view x_i as $v_1 h_{i+1,1}$. Then Mahowald's answer would follow from the differentials

- $d_3 h_{i1} = v_1^{-2} h_{11} h_{21} h_{i-1,1}^2$ (for $i \geq 3$),
- $d_3 v_1^2 = h_{11}^3$,

together with some sort of multiplicativity properties. Unfortunately, we do not know how to produce these differentials, and there is not much apparent multiplicative structure in the spectral sequence. We know the following.

(a) The Q_1-based spectral sequence converging to $v_1^{-1}\pi_{**}M$ is a module over the Q_1-based spectral sequence for the sphere, which has

$$E_2 \cong \mathbf{F}_2[v_1^{\pm 1}] \otimes \mathbf{F}_2[h_{10}, h_{11}, h_{21}, h_{31}, \dots].$$

The latter spectral sequence need not converge, and indeed it is not even clear what the abutment is, but it is a spectral sequence of algebras, and the comodule map $\mathbf{F}_2 \to M$ induces a surjection at E_2. Hence if one can compute any differentials for the sphere spectral sequence, that has implications for the Moore spectrum.

(b) In the abutment of the spectral sequence for the Moore spectrum, $h_{11}^3 = 0$.

(c) Davis and Mahowald [**DM88**, Theorem 1.3] have computed the v_1-inverted Ext of the spectrum $Y = H_*(\mathbf{R}P^2 \wedge \mathbf{C}P^2) = \mathbf{F}_2[x_1]/(x_1^4)$. They found that
$$v_1^{-1} \operatorname{Ext}_A^{**}(\mathbf{F}_2, H_*Y) \cong \mathbf{F}_2[v_1^{\pm 1}] \otimes \mathbf{F}_2[h_{21}, h_{31}, h_{41}, \dots].$$
As Eisen points out in [**Eis87**], this follows from his computation; from our point of view, the Q_1-based Adams spectral sequence for Y embeds in the Q_1-based Adams spectral sequence for $Y(1)$, and hence collapses. The cofibration $M \to Y \to \Sigma^{0,2} M$ in $\mathsf{Stable}(A)$, corresponding to the Ext class h_{11}, leads to a Bockstein spectral sequence with E_1-term the same as the E_2-term for the Q_1-based spectral sequence converging to $v_1^{-1}\pi_{**}M$. One might be able to use this to aid in the computation of differentials in that spectral sequence.

We end this section by discussing the odd primary case. When p is odd, Miller [**Mil78**, Corollary 3.6] calculated $v_1^{-1} \operatorname{Ext}_A^{**}(\mathbf{F}_p, H_*X)$, where X is the mod p Moore spectrum:
$$v_1^{-1} \operatorname{Ext}_A^{**}(\mathbf{F}_p, H_*X) \cong \mathbf{F}_p[v_1^{\pm 1}] \otimes \Lambda[h_{t0} : t \geq 1] \otimes \mathbf{F}_p[b_{t0} : t \geq 1].$$
This is the E_2-term of the Q_1-based Adams spectral sequence. The spectral sequence collapses in this case, because of two things: first, since the mod p Moore spectrum is a ring spectrum, this is a spectral sequence of algebras. Second, for easy degree reasons, the elements h_{t0} and b_{t0} are permanent cycles; a more careful analysis shows that v_1^n is a permanent cycle for all n—each differential d_r takes v_1^n to a tridegree which is zero. In a spectral sequence of algebras in which each algebra generator is a permanent cycle, the whole spectral sequence consists of permanent cycles.

CHAPTER 5

Quillen stratification and nilpotence

The vanishing line theorem of Section 2.3 is a nilpotence theorem of a sort: if X is a finite spectrum, then any self-map of X with slope smaller than that of the vanishing line for X must be nilpotent (because some power of it will lie above the vanishing line, where "above" is in terms of the Adams spectral sequence grading of Figure 2.3.A). This vanishing line theorem has applications; for example, we used it in Section 2.4 to construct a non-nilpotent self-map of any finite spectrum.

In this chapter we give some related, but stronger, results; we work mostly at the prime 2. We let $\mathscr{Q}$ denote the category of quasi-elementary quotients of A, with morphisms given by quotient maps. We can assemble the individual Hurewicz (restriction) maps $S^0 \to HE$ into a homotopy limit

$$S^0 \to \varprojlim_{\mathscr{Q}} HE.$$

Since the maximal quasi-elementary quotients are conormal, there is an action of A on the $\mathbf{F}_p$-algebra $\varprojlim_{\mathscr{Q}} HE_{**}$; we use a superscript A to denote the invariants under this action (see Definition 5.1.1). We prove that

$$\pi_{**} S^0 \to (\varprojlim_{\mathscr{Q}} HE_{**})^A$$

is an F-isomorphism—its kernel consists of nilpotent elements, and some pth power of every element in the target is actually in the image. One can view this as an analogue of the Quillen stratification theorem [**Qui71**, Theorem 6.2], which identifies $\pi_{**} S^0$ up to F-isomorphism in the category $\mathsf{Stable}(kG^*)$, for G a finite group and k an algebraically closed field of characteristic p.

Similarly, suppose that we write D for the following quotient of A:

$$D = A/(\xi_1^p, \xi_2^{p^2}, \xi_3^{p^3}, \ldots, \xi_n^{p^n}, \ldots).$$

Then D is conormal, and when $p = 2$, the Hurewicz map induces an F-isomorphism

$$\pi_{**} S^0 \to (HD_{**})^A.$$

These F-isomorphism theorems first appeared in [**Pal99**].

Quillen's theorem about group cohomology with trivial coefficients was generalized to nontrivial coefficients by Avrunin and Scott in [**AS82**]; their result suggests that a direct analogue of these F-isomorphism theorems with nontrivial coefficients—for example, that $[X, X]$ is F-isomorphic to $(\varprojlim_{\mathscr{Q}} [X, HE \wedge X]_{**})^A$—is too much to hope for. In this chapter, we give several weaker results with nontrivial coefficients: if R is a ring spectrum, then an element $\alpha \in \pi_{**} R$ is nilpotent if and only if its image under the Hurewicz map in $HD_{**} R$ is nilpotent, or if its image in $HE_{**} R$ is zero for all quasi-elementary quotients E of A. There are similar results for detecting the nilpotence of self-maps of finite spectra. (These theorems

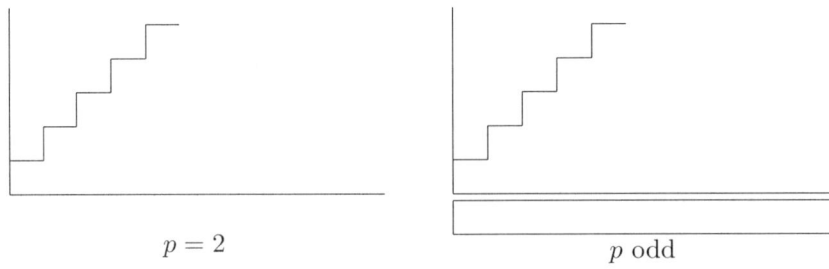

FIGURE 5.1.A. Profile functions for $D = A/(\xi_1^p, \ldots, \xi_n^{p^n}, \ldots)$.

are improvements on results of the author in [**Pal96a**].) In the next chapter, we discuss this further: see Theorem 6.2.3 for a statement about lifting A-invariant elements from $HD_{**}R$ to $\pi_{**}R$, and Conjecture 6.7.1 for the best generalization of the F-isomorphism theorem we could hope for.

We state our results more precisely in Section 5.1. In Section 5.2, we prove the two theorems—nilpotence and F-isomorphism—involving the quotient Hopf algebra D; we then use these to prove the analogous results for quasi-elementary Hopf algebras in Section 5.3. We end the chapter with Sections 5.4 and 5.5. In the first of these, we indicate why our proofs only work at the prime 2; this leads us to a conjecture about the nilpotence of particular classes in Ext over quotients of A. In the second, we discuss a few possible generalizations of the main results of this chapter.

Except for Sections 5.4 and 5.5, we work at the prime 2 in this chapter.

5.1. Statements of theorems

Let $p = 2$. In this section we present two F-isomorphism results and two nilpotence results.

Let D be the following quotient Hopf algebra of A:

$$D = A/(\xi_1^2, \xi_2^4, \xi_3^8, \ldots, \xi_n^{2^n}, \ldots).$$

See Figure 5.1.A for its profile function (along with the analogous picture at odd primes). Note that D and A have the same quasi-elementary quotients—i.e., every quasi-elementary quotient map $A \twoheadrightarrow E$ factors as $A \twoheadrightarrow D \twoheadrightarrow E$. As a result, it turns out that HD plays a similar role in $\mathsf{Stable}(A)$ to that of BP in the ordinary stable homotopy category, at least as far as detection of nilpotence goes. We write $\eta_D \colon S^0 = HA \to HD$ for the unit map of the ring spectrum HD, and similarly for $\eta_E \colon S^0 \to HE$ for E quasi-elementary.

5.1.1. Quillen stratification. In this subsection, we state our results describing $\pi_{**}S^0$ up to F-isomorphism. The results here first appeared in [**Pal99**].

DEFINITION 5.1.1.   (a) Given a Hopf algebra B and a B-comodule M, we define the B-*invariants* of M to be

$$M^B = \mathrm{Hom}_B(\mathbf{F}_p, M) \subseteq M.$$

If B is connected, this is the same as the *primitives* PM of M: if $\psi \colon M \to B \otimes M$ is the coaction map, then we let $PM = \{m \in M \mid \psi(m) = 1 \otimes m\}$.

(b) Following Quillen [**Qui71**], if k is a field of characteristic $p > 0$ and if $\varphi \colon R \to S$ is a map of graded commutative k-algebras, we say that φ is an *F-isomorphism* if it satisfies the following properties.
  (i) Every $x \in \ker \varphi$ is nilpotent. (Hence $\sqrt{\ker \varphi}$ is the nilradical of R.)
  (ii) For any element $y \in S$, there is an integer n so that $y^{p^n} \in \operatorname{im} \varphi$.
In (ii), if one can choose the same n for every y, then we say that φ is a *uniform F-isomorphism*.

The definition of F-isomorphism is intended to capture the properties of the Frobenius map
$$F \colon \mathbf{F}_p[x_1, \ldots, x_m] \to \mathbf{F}_p[x_1, \ldots, x_m]$$
and its iterates F^n, where F is defined by $F(\alpha) = \alpha^p$. There is another notion, called "N-isomorphism," in which condition (ii) is replaced by

  (ii)' For any element $y \in S$, there is an integer n so that $y^n \in \operatorname{im} \varphi$.

When working in characteristic p, F-isomorphism is a more natural concept. It is also stronger than N-isomorphism, and it is what our results produce.

Here is our first result. Theorem 2.1.1 tells us that D is conormal, so by Remark 1.2.13, there is a coaction of $A \square_D \mathbf{F}_2$ on HD_{**}.

THEOREM 5.1.2 (Quillen stratification, I). *The Hurewicz map $\pi_{**} S^0 \to HD_{**}$ factors through*
$$\varphi \colon \pi_{**} S^0 \to HD_{**}^{A \square_D \mathbf{F}_2},$$
and φ is an F-isomorphism.

Here is our second result, an analogue of Quillen's theorem [**Qui71**, Theorem 6.2], which identifies group cohomology up to F-isomorphism. We have defined in Definition 2.1.10 (see also Proposition 2.1.12) the notion of a quasi-elementary quotient Hopf algebra of A; we let $\mathscr{Q}$ denote the category of quasi-elementary quotients of A, with morphisms given by quotient maps. In Section 5.3 below, we construct a coaction of A on $\varprojlim_{\mathscr{Q}} HE_{**}$.

THEOREM 5.1.3 (Quillen stratification, II). *The map $\pi_{**} S^0 \to \varprojlim_{\mathscr{Q}} HE_{**}$ factors through*
$$\gamma \colon \pi_{**} S^0 \to \left(\varprojlim_{\mathscr{Q}} HE_{**} \right)^A,$$
and γ is an F-isomorphism.

Both inverse limits here are inverse limits of $\mathbf{F}_2$-algebras.

We prove Theorem 5.1.2 in Section 5.2; we show that Theorem 5.1.3 follows from Theorem 5.1.2 in Section 5.3.

Lin proved special cases of Theorems 5.1.2 and 5.1.3 in [**Lin77a**, **Lin77b**]; his results applied not to $\operatorname{Ext}_A^{**}(\mathbf{F}_2, \mathbf{F}_2)$ but to $\operatorname{Ext}_B^{**}(\mathbf{F}_2, \mathbf{F}_2)$, where B is any quotient Hopf algebra of A with a finite profile function: for each n, $\xi_n^{2^i} = 0$ in B for some i. In this setting, it turns out that some power of every cohomology element is invariant, so one does not have to apply $(-)^A$. There are also weaker forms of our results in [**Wil81**], [**HP93**], and [**Pal96a**]; these papers all focus on the detection of nilpotence (finding a map f that satisfies condition (i) in Definition 5.1.1(b)), rather than on the identification of all of the non-nilpotent elements.

REMARK 5.1.4. (a) One can view Theorem 5.1.3 as giving an analogue of the Quillen stratification theorem [**Qui71**, Theorem 6.2]: given a finite group G and an algebraically closed field k of characteristic p, we let $\mathscr{A}$ be the category whose objects are the elementary abelian p-subgroups of G, and whose morphisms are generated by inclusions and compositions. Then the natural map

$$H^*(G;k) \to \varprojlim_{E \in \mathscr{A}} H^*(E;k)$$

is an F-isomorphism. The role of the conjugation maps in the category $\mathscr{A}$ is played, in our results, by the process of taking invariants.

(b) One can also view Theorem 5.1.2 as an analogue of Nishida's theorem [**Nis73**]—in the ordinary stable homotopy category, $\pi_* S^0$ is isomorphic to **Z**, mod nilpotence; however, while Nishida's theorem does not immediately lead to guesses as to further structure in the ordinary stable homotopy category, our analogue does. For instance, see Section 6.7 for a suggested classification of thick subcategories of finite spectra in Stable(A).

(c) We have an explicit formula for the algebra $\varprojlim HE_{**}$ (Proposition 6.5.1), as well as for the coaction of A on $\varprojlim HE_{**}$ (Proposition 6.5.6). One can use this to predict the presence of large families of non-nilpotent elements in $\pi_{**} S^0 = \mathrm{Ext}_A^{**}(\mathbf{F}_2, \mathbf{F}_2)$. See Section 6.5 for details.

(d) We do not expect the F-isomorphisms in these results to be uniform, although we do not have much evidence either way.

(e) Of course, if A' is any quotient Hopf algebra of A, then similar results hold: for instance, the map $HA'_{**} \to (HD'_{**})^{A'}$ is an F-isomorphism, where D' is the apparent quotient of A'. The proofs for Theorems 5.1.3 and 5.1.2 carry over easily.

5.1.2. Nilpotence. We move on to our nilpotence theorems. The first of these is based on the nilpotence theorem of Devinatz, Hopkins, and Smith [**DHS88**]: the ring spectrum BP detects nilpotence. Note that if E is a ring spectrum and Y is any spectrum, then $[Y, E \wedge Y]_{**}$ is a ring: given $f, g \in [Y, E \wedge Y]_{**}$, define the product fg to be the composite

$$Y \xrightarrow{g} E \wedge Y \xrightarrow{1_E \wedge f} E \wedge E \wedge Y \xrightarrow{\mu \wedge 1_Y} E \wedge Y,$$

where μ is the multiplication map on E. Also, we say that a map of spectra $f \colon F \to X$ is *smash nilpotent* if some smash power of f, $f^{\wedge n} \colon F^{\wedge n} \to X^{\wedge n}$, is null.

THEOREM 5.1.5 (Nilpotence theorem, I). *The ring spectrum HD detects nilpotence*:

(a) *Fix a ring spectrum R and $\alpha \in \pi_{**} R$. Then α is nilpotent if and only if $HD_{**}(\alpha^n) = 0$ for some n.*

(b) *Fix a finite spectrum Y and a self-map $f \colon Y \to Y$. Then $f \in [Y, Y]_{**}$ is nilpotent if and only if $HD_{**} f$ is nilpotent.*

(c) *Fix a finite spectrum F, an arbitrary spectrum X, and a map $f \colon F \to X$. Then $f \in [F, X]_{**}$ is smash-nilpotent if $1_{HD} \wedge f \in [HD \wedge F, HD \wedge X]_{**}$ is zero.*

The corresponding result for modules is [**Pal96a**, Theorems 3.1 and 4.2]. In [**Pal96a**] we prove this in enough generality so that the proof goes through here without difficulty. We also give a (slightly different) proof in Section 5.2.

The second nilpotence theorem is, more or less, an analogue of the $K(n)$ nilpotence theorem in [**HS98**]. The following appeared for bounded below modules in [**Pal96a**, Theorems 1.1 and 4.3].

THEOREM 5.1.6 (Nilpotence theorem, II). *The collection of ring spectra*
$$\{HE \mid E \text{ quasi-elementary}\}$$
detects nilpotence:
  (a) *Fix a ring spectrum R and $\alpha \in \pi_{**}R$. Then α is nilpotent if $HE_{**}\alpha = 0$ is zero for all quasi-elementary quotients E of A.*
  (b) *Fix a finite spectrum Y and a self-map $f \colon Y \to Y$. Then f is nilpotent if and only if $HE_{**}f$ is nilpotent for all quasi-elementary quotients E of A.*
  (c) *Fix a finite spectrum F, an arbitrary spectrum X, and a map $f \colon F \to X$. Then f is smash-nilpotent if $1_E \wedge f \colon HE \wedge F \to HE \wedge X$ is zero for all quasi-elementary quotients E of A.*

One might hope to replace part (a) with an "if and only if" statement, but since R need not be finite, and since there are infinitely many quasi-elementary quotients of A, this may be problematic (compare to Theorem 5.1.5(a), in which there is a single detecting ring spectrum). For part (b), on the other hand, the finiteness of Y essentially means that if $\eta_E \wedge f$ is nilpotent for all E, then some iterate of this map is zero for all E. All three parts have an "if and only if" version in ordinary stable homotopy theory (see [**HS98**, Theorem 3]). The proof of that theorem relies on nice properties of the Morava K-theories, properties that are not shared by the objects HE in our setting; in particular, the $K(n)$'s are field spectra, and the HE's are not. Hence we have a slightly weaker theorem than Hopkins and Smith do.

The proof of the analogous result in [**Pal96a**] does not apply to nonconnective situations, so we give a new proof below.

REMARK 5.1.7. (a) Suppose that F and R are ring spectra and fix $\alpha \in \pi_{**}R$. Then $F_{**}\alpha = 0$ if and only if $\eta_F \wedge \alpha \colon S^0 \to F \wedge R$ is null. So one may replace the hypothesis in Theorems 5.1.5 and 5.1.6 that $HD_{**}(\alpha^n) = 0$ (resp., $HE_{**}\alpha = 0$) with the condition that $\eta_D \wedge \alpha^n$ is null (resp., that $\eta_E \wedge \alpha$ is null).
  (b) Note that $\mathbf{F}_2$ is a quasi-elementary quotient Hopf algebra of A, so $H\mathbf{F}_2$ is included as one of the detecting spectra in Theorem 5.1.6. (Compare with [**HS98**], where $H\mathbf{F}_p = K(\infty)$ is included for parts (a) and (c), but not for (b)).
  (c) Fix a quotient Hopf algebra A' of A, and let D' be the quotient of A' induced by $A \twoheadrightarrow D$. One can generalize Theorem 5.1.5 in an obvious sort of way: if Y is a finite spectrum and $f \colon Y \to HA' \wedge Y$ is a "self-map," then
$$f \in [Y, HA' \wedge Y]_{**} \text{ is nilpotent}$$
$$\Leftrightarrow \eta_{D'} \wedge f \in [Y, HD' \wedge Y]_{**} \text{ is nilpotent}$$
$$\Leftrightarrow \eta_{E'} \wedge f \in [Y, HE' \wedge Y]_{**} \text{ is nilpotent for all } E'.$$

(Here E' ranges over all quasi-elementary quotients of A'.) There are similar versions of the ring spectrum and smash-nilpotence results. The proofs are straightforward generalizations of the ones below, so we omit them.

(d) The quotient Hopf algebra D is "best possible," in the sense that if B is a quotient of A which does not map onto D, then there are non-nilpotent elements in $\pi_{**}S^0$ which are in the kernel of $\pi_{**}S^0 \to HB_{**}$. One can see this from Theorem 5.1.2 and Example 6.5.7(a).

(e) Theorem 5.1.5 identifies a single ring spectrum HD which detects nilpotence, but we do not know its coefficient ring completely. See Propositions 5.3.4 and 6.5.1 for partial information. On the other hand, we have computed the coefficient rings of HE for E quasi-elementary in Proposition 2.1.9—they are polynomial rings.

(f) One does not need to use all of the quasi-elementary quotient Hopf algebras of A to detect nilpotence; for instance, one can use only the maximal quasi-elementary quotients, or only the finite-dimensional ones.

(g) In fact, one can see from the proof of Theorem 5.1.6 that one only needs the following spectra to detect nilpotence:
$$\{h_{m+1,j}^{-1} H(E(m)/(\xi_{m+1}^{2^{j+1}})) \mid m \geq 0,\ 0 \leq j \leq m\}.$$
Here $E(m)$ is one of the maximal quasi-elementary quotients of A, defined in Corollary 2.1.8, and $E(m)/(\xi_{m+1}^{2^{j+1}})$ is a further quotient of that. By Lemmas 1.2.15 and 2.1.2, $H(E(m)/(\xi_{m+1}^{2^{j+1}}))$ has a self-map $h_{m+1,j}$, and
$$h_{m+1,j}^{-1} H(E(m)/(\xi_{m+1}^{2^{j+1}}))$$
denotes the mapping telescope of this self-map.

5.2. Nilpotence and F-isomorphism via the Hopf algebra D

In this section we show that the spectrum HD detects a lot of information: we prove that it detects $\pi_{**}S^0$ modulo nilpotent elements (Theorem 5.1.2), and that it detects non-nilpotent elements of $\pi_{**}R$ for any ring spectrum R (Theorem 5.1.5).

The Hopf algebra D is a conormal quotient of A, by Theorem 2.1.1. It is standard that for conormal quotient Hopf algebras, the restriction (Hurewicz) map $h\colon \pi_{**}S^0 \to HD_{**}$ factors as advertised in Theorem 5.1.2:
$$\varphi\colon \pi_{**}S^0 \to HD_{**}^{A \square_D \mathbf{F}_2}.$$
One way to see this is to consider the Hopf algebra extension
$$A \square_D \mathbf{F}_2 \to A \to D,$$
and the associated spectral sequence (as in Section 1.4), which has
$$E_2^{s,t} = \mathrm{Ext}_{A \square_D \mathbf{F}_2}^s(\mathbf{F}_2, \mathrm{Ext}_D^t(\mathbf{F}_2, \mathbf{F}_2)) \Rightarrow \mathrm{Ext}_A^{s+t}(\mathbf{F}_2, \mathbf{F}_2).$$
The desired factorization is just the edge homomorphism
$$\mathrm{Ext}_A^t(\mathbf{F}_2, \mathbf{F}_2) \to E_2^{0,t}.$$
We have to prove Theorem 5.1.5—the spectrum HD detects nilpotence—and to verify conditions (i) and (ii) of Definition 5.1.1(b)—every element of $\ker \varphi$ is nilpotent, and for every element $y \in HD_{**}^{A \square_D \mathbf{F}_2}$, there is an integer m so that $y^{2^m} \in \mathrm{im}\,\varphi$.

Note that condition (i) follows from Theorem 5.1.5(a) with $R = S^0$. Also, the proof of Theorem 5.1.5 and the verification of (ii) are quite similar, so first we lay the groundwork for both.

For each integer $n \geq 1$, we let
$$D(n) = A/(\xi_1^2, \xi_2^4, \ldots, \xi_n^{2^n}).$$

5.2. NILPOTENCE AND F-ISOMORPHISM VIA THE HOPF ALGEBRA D

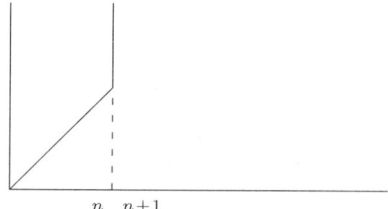

FIGURE 5.2.A. Profile function for $D(n)$. As in Figure 2.1.B, the diagonal line is an abbreviation for a staircase shape.

We let $D(0) = A$. See Figure 5.2.A for the profile function. Each $D(n)$ is a conormal quotient Hopf algebra of A, and we have a diagram of Hopf algebra surjections:
$$A = D(0) \twoheadrightarrow D(1) \twoheadrightarrow D(2) \twoheadrightarrow \cdots .$$
D is the colimit of this diagram.

First we discuss how to lift information from HD_{**} to $HD(n)_{**}$ for some n. We have the following lemma.

LEMMA 5.2.1. (a) *We have* $HD_{**} = \varinjlim HD(n)_{**}$. *More generally, if X is any connective spectrum, then $HD_{**}X = \varinjlim HD(n)_{**}X$, and this colimit stabilizes in each bidegree.*

(b) *We have* $HD_{**}^{A \square_D \mathbf{F}_2} = \varinjlim \left(HD(n)_{**}^{A \square_{D(n)} \mathbf{F}_2} \right)$.

(These colimits are of $\mathbf{F}_2$-vector spaces.)

PROOF. Since homotopy commutes with direct limits, the first sentence of part (a) is clear. The second follows from an easy connectivity argument, as in the proof of Lemma 2.4.4.

Part (b): The coaction of A on $HD(n)_{**}$ is defined in Remark 1.2.13. Note that each restriction map $HD(n)_{**} \to HD(n+1)_{**}$ is an A-comodule map (and in fact, a map of comodule algebras). Applying $\text{Hom}_A^*(\mathbf{F}_2, -)$ gives the following:
$$\text{Hom}_A(\mathbf{F}_2, HD_{**}) = \text{Hom}_A(\mathbf{F}_2, \varinjlim HD(n)_{**}) = \varinjlim \text{Hom}_A(\mathbf{F}_2, HD(n)_{**}),$$
where the last equality holds because $\mathbf{F}_2$ is small in the category of A-comodules. Since $\text{Hom}_A(\mathbf{F}_2, HD(n)_{**}) = HD(n)_{**}^A = HD(n)_{**}^{A \square_{D(n)} \mathbf{F}_2}$, we have the desired result. □

Now we discuss how to take information about $HD(n)_{**}$ and get information about $HD(n-1)_{**}$. We want to know about $\pi_{**}S^0 = HD(0)_{**}$, so we will eventually want to use downward induction on n.

Not only is each $D(n)$ a conormal quotient of A, it is also a conormal quotient of $D(n-1)$. The Hopf algebra kernel of the quotient map is easy to identify: there is an extension of Hopf algebras
$$\mathbf{F}_2[\xi_n^{2^n}] \to D(n-1) \to D(n),$$
where $\xi_n^{2^n}$ is primitive in the kernel. So given any $D(n-1)$-comodule M, there is a change-of-rings spectral sequence (see (1.4.8)) with
$$E_2^{s,t,u}(M) = \text{Ext}_{\mathbf{F}_2[\xi_n^{2^n}]}^{s,u}(\mathbf{F}_2, \text{Ext}_{D(n)}^{t,*}(\mathbf{F}_2, M)) \Rightarrow \text{Ext}_{D(n-1)}^{s+t,u}(\mathbf{F}_2, M).$$

By Proposition 1.4.9, in the category Stable($D(n-1)$) of cochain complexes of injective left $D(n-1)$-comodules, this spectral sequence is the same, up to regrading, as the $HD(n)$-based Adams spectral sequence. Corollary 1.4.7 extends this from the case when M is a comodule to an arbitrary object X in Stable($D(n-1)$). Because of this, for parts of the proof we will work in the category Stable($D(n-1)$). We also use the grading on the spectral sequence as given here; we do *not* use the Adams spectral sequence grading from Section 1.4. Throughout, we abuse notation somewhat, writing $\operatorname{Ext}^{**}_{D(n)}(\mathbf{F}_2, X)$ for $\pi_{**}(HD(n) \wedge X)$.

We need to establish an important property of this spectral sequence, that it has a nice vanishing plane at some E_r-term. We write $E_r^{s,t,u}(X)$ for the spectral sequence converging to $\operatorname{Ext}^{**}_{D(n-1)}(\mathbf{F}_2, X) = HD(n-1)_{**}X$.

PROPOSITION 5.2.2. *Fix a connective spectrum X and an integer $m \neq 1$. For some numbers r and c, $E_r^{s,t,u}(X) = 0$ when $ms + t - u > c$.*

We prove this using Theorem 1.4.4, which says that vanishing planes in Adams spectral sequences are generic. (And again, this is an Adams spectral sequence, as long as we work in the category Stable($D(n-1)$).)

(As remarked above, the change-of-rings spectral sequence E_r^{***} is the same as the Adams spectral sequence, up to a regrading. Writing the Adams spectral sequence as ${}^{\natural}E_r^{***}$, then with m as in the proposition, the vanishing plane for ${}^{\natural}E_r^{***}$ has the following form:
$$ {}^{\natural}E_2^{s,t,u}(W \wedge X) = 0 $$
when $ms - t > c$, i.e., when
$$ s \geq \frac{-1}{m-1}(s+u) + \frac{1}{m-1}(t+u) + \frac{c}{m-1}. $$
In particular, if $m \neq 1$, we can apply Theorem 1.4.4.)

We fix a connective spectrum X. "(i_0, j_0)-connectivity" is defined in Definition 1.3.4.

LEMMA 5.2.3. *For each integer m, there is a finite $D(n-1)$-comodule W so that*
$$ E_2^{s,t,u}(W \wedge X) = 0 $$
when $ms + t - u > c$, where c depends only on the connectivity of X. More precisely, if X is (i_0, j_0)-connective, then
$$ E_2^{s,t,u}(W \wedge X) = 0 $$
when $ms + t - u > -j_0$.

PROOF. Choose an integer $j \geq n$ so that $|\xi_n^{2^j}| = 2^j(2^n - 1) > m$, and let $W = W_j = \mathbf{F}_2[\xi_n^{2^n}]/(\xi_n^{2^j})$, with the apparent $D(n-1) \square_{D(n)} \mathbf{F}_2$-comodule structure. Then W_j is a trivial $D(n)$-comodule (by definition), and as coalgebras we have
$$ \mathbf{F}_2[\xi_n^{2^n}] \cong W_j \otimes \mathbf{F}_2[\xi_n^{2^j}]. $$
In particular, $W_j = \mathbf{F}_2[\xi_n^{2^n}] \square_{\mathbf{F}_2[\xi_n^{2^j}]} \mathbf{F}_2$, and for any $\mathbf{F}_2[\xi_n^{2^n}]$-comodule N there is a change-of-rings isomorphism (as in Corollary 1.2.7—see also Example 1.2.4 and Lemma 1.2.5):
$$ \operatorname{Ext}^{**}_{\mathbf{F}_2[\xi_n^{2^n}]}(\mathbf{F}_2, W_j \otimes N) \cong \operatorname{Ext}^{**}_{\mathbf{F}_2[\xi_n^{2^j}]}(\mathbf{F}_2, N). $$

Hence the E_2-term of the spectral sequence for $W_j \wedge X$ looks like
$$E_2^{s,t,u} \cong \mathrm{Ext}^{s,u}_{\mathbf{F}_2[\xi_n^{2^n}]}(\mathbf{F}_2, \mathrm{Ext}^{t,*}_{D(n)}(\mathbf{F}_2, W_j \wedge X))$$
$$\cong \mathrm{Ext}^{s,u}_{\mathbf{F}_2[\xi_n^{2^n}]}(\mathbf{F}_2, W_j \otimes \mathrm{Ext}^{t,*}_{D(n)}(\mathbf{F}_2, X))$$
$$\cong \mathrm{Ext}^{s,u}_{\mathbf{F}_2[\xi_n^{2^j}]}(\mathbf{F}_2, \mathrm{Ext}^{t,*}_{D(n)}(\mathbf{F}_2, X)).$$

The Hopf algebra $\mathbf{F}_2[\xi_n^{2^j}]$ is $|\xi_n^{2^j}|$-connected, so if L is a comodule which is zero below degree t, then
$$\mathrm{Ext}^{s,u}_{\mathbf{F}_2[\xi_n^{2^j}]}(\mathbf{F}_2, L) = 0$$
when $u < |\xi_n^{2^j}|s + t$. Now note that if X is (i_0, j_0)-connective, then $HD(n)_{t,*}X = \mathrm{Ext}^{t,*}_{D(n)}(\mathbf{F}_2, X)$ is zero below degree $t + j_0$. □

Next we need to show that X is in thick$(W_j \wedge X)$, in which case genericity of vanishing planes will yield one for X. We start with the following lemma, which describes how to build W_j out of $\mathbf{F}_2$ in a nice way.

LEMMA 5.2.4. *For $j \geq n$, let $W_j = \mathbf{F}_2[\xi_n^{2^n}]/(\xi_n^{2^j})$. Then there is a short exact sequence of $D(n-1) \square_{D(n)} \mathbf{F}_2$-comodules*
$$0 \to W_j \to W_{j+1} \to s^{|\xi_n^{2^j}|} W_j \to 0.$$
(As in Section 0.1, s denotes the suspension functor in the category of graded comodules.) The connecting homomorphism in $\mathrm{Ext}^{**}_{D(n-1)}(\mathbf{F}_2, -)$ is multiplication by
$$h_{nj} = [\xi_n^{2^j}] \in \mathrm{Ext}^{**}_{D(n-1)}(\mathbf{F}_2, \mathbf{F}_2).$$
Replacing W_j by its injective resolution gives a cofibration sequence
$$\Sigma^{1, |\xi_n^{2^j}|} W_j \xrightarrow{h_{nj}} W_j \to W_{j+1} \to \Sigma^{0, |\xi_n^{2^j}|} W_j.$$

(We are abusing notation a bit here, by writing W_j for both the module and its injective resolution. We will continue this practice for the remainder of this section.)

PROOF. This follows from Lemma 1.2.15. □

LEMMA 5.2.5. *Let Y and Z be spectra. Given a cofiber sequence $Y \xrightarrow{f} Y \to Z$, if the map f is nilpotent under composition, then $Y \in \mathrm{thick}(Z)$.*

(As usual, we are omitting suspensions. The self-map f could have nonzero degree here.)

PROOF. Write $C(f^i)$ for the cofiber of f^i. First we show by induction on i that $C(f^i)$ is in thick(Z) for each $i \geq 1$. The $i = 1$ case is immediate, because $C(f) = Z$. Now assume that $C(f^{i-1}) \in \mathrm{thick}(Z)$, and consider the following diagram, in which each row and column is a cofibration (cf. [**HS98**, Lemma 2.6]):

$$\begin{array}{ccccc} Y & \xrightarrow{f^{i-1}} & Y & \longrightarrow & C(f^{i-1}) \\ \parallel & & \downarrow f & & \downarrow \\ Y & \xrightarrow{f^i} & Y & \longrightarrow & C(f^i) \\ \downarrow & & \downarrow & & \downarrow \\ * & \longrightarrow & Z & = & Z. \end{array}$$

The right-hand column shows that $C(f^i)$ is in thick(Z), as desired.

If $f^n = 0$, then $C(f^n)$ splits as $C(f^n) = Y \vee Y$. Since $C(f^n) \in$ thick(Z), so is Y. □

Suppose that $j \geq n$; then a result of Lin [**Lin77a**], stated as Theorem B.2.1(a) below, tells us that the element $h_{nj} \in HD(n-1)_{**}$ is nilpotent. Hence we have the following:

LEMMA 5.2.6. *If $j \geq n$, then $X \in$ thick($W_j \wedge X$).*

PROOF. We show by downward induction on i that for $j \geq i \geq n$, $W_i \wedge X$ is in thick($W_j \wedge X$); the lemma is proved when $i = n$, since $W_n = S^0$.

The induction starts (trivially) with $i = j$. Suppose that $i < j$, and consider the cofibration sequence in Lemma 5.2.4. Because h_{ni} is nilpotent, Lemma 5.2.5, applied with $Y = W_i \wedge X$ and $Z = W_{i+1} \wedge X$, shows that $W_i \wedge X \in$ thick($W_{i+1} \wedge X$), and hence is in thick($W_j \wedge X$) by induction. □

PROOF OF PROPOSITION 5.2.2. This follows immediately from Lemma 5.2.3, Lemma 5.2.6, and the genericity of vanishing planes, Theorem 1.4.4. □

5.2.1. Nilpotence: Proof of Theorem 5.1.5. Now we prove Theorem 5.1.5, and hence verify condition (i) of Definition 5.1.1(b). Recall from Remark 5.1.7(a) that if R is a ring, then for $\alpha \in \pi_{**}R$, $HD_{**}(\alpha^n) = 0$ if and only if $\eta_D \wedge \alpha^n = 0$. Since $\eta_D \colon S^0 \to HD$ is the unit of the ring spectrum HD, then in $\pi_{**}HD$, $\eta_D^n = \eta_D$ for all n; so $\eta_D \wedge \alpha^n = 0$ if and only if $\eta_D^n \wedge \alpha^n = 0$.

PROOF OF THEOREM 5.1.5. The basic idea of the proof is, of course, based on that of the nilpotence theorem in [**DHS88**]. As in that proof, one can reduce to the ring spectrum case—part (a)—in which the ring is connective. So we let R be a connective ring spectrum, we fix $\alpha \in \pi_{**}R$ and assume that $\eta_D \wedge \alpha$ is nilpotent. By raising α to a power, we may assume that $\eta_D \wedge \alpha = 0$. We want to show that α is nilpotent in $\pi_{**}R = HD(0)_{**}R$.

By Lemma 5.2.1, since $\eta_D \wedge \alpha = 0$, then we must have $\eta_{D(n)} \wedge \alpha = 0$ for some n. We want to show that if $\eta_{D(n)} \wedge \alpha = 0$, then $\eta_{D(n-1)} \wedge \alpha^j = 0$ for some j; the result will follow by downward induction on n.

Consider the change-of-rings spectral sequence

$$E_2^{s,t,u}(R) = \mathrm{Ext}^{s,u}_{\mathbf{F}_2[\xi_n^{2^n}]}(\mathbf{F}_2, \mathrm{Ext}^{t,*}_{D(n)}(\mathbf{F}_2, R)) \Rightarrow \mathrm{Ext}^{s+t,u}_{D(n-1)}(\mathbf{F}_2, R).$$

Write z for $\eta_{D(n-1)} \wedge \alpha$. Since $\eta_{D(n)} \wedge \alpha = 0$, then z must be represented by a class $\tilde{z} \in E_2^{s,t,u}$ with $s > 0$.

We can find an integer $m \neq -1$ so that $ms + t - u > 0$, and then for any c, we can find a j so that
$$msj + tj - uj > c.$$
By Proposition 5.2.2 (with $X = R$), for some r and c we have $E_r^{s,t,u} = 0$ when $ms + t - u > c$. As noted above, we can choose j so that $\tilde{z}^j$ lies above this vanishing plane, so at the E_r-term for which we have the vanishing plane, $\tilde{z}^j$ must be zero. Thus $\tilde{z}^j$ is zero at E_∞, and hence is zero in the abutment modulo terms of higher filtration. The higher filtration pieces are also above the vanishing plane, though, and are therefore zero. So $z^j = \eta_{D(n-1)}^j \wedge \alpha^j = 0$ in the abutment, $HD(n-1)_{**}R$; since $\eta_{D(n-1)}^j = \eta_{D(n-1)}$, we are done. □

5.2.2. F-isomorphism: Proof of Theorem 5.1.2.
We need to show that the map
$$\varphi \colon \pi_{**}S^0 \to HD_{**}^{A\,\square_D\,\mathbf{F}_2}$$
is an F-isomorphism. By Theorem 5.1.5, it satisfies part (i) of Definition 5.1.1(b); i.e., it is a monomorphism mod nilpotents. Now we have to verify condition (ii).

VERIFICATION OF CONDITION (ii). Fix $y \in HD_{i,j}^{A\,\square_D\,\mathbf{F}_2}$. We want to show that there is an integer m so that $y^{2^m} \in \operatorname{im}\varphi$. By Lemma 5.2.1(b), there is an n so that y lifts to $HD(n)_{**}^{A\,\square_{D(n)}\,\mathbf{F}_2}$. (Alternatively, one can use Lemma 5.2.1(a) to lift y to $HD(n)_{**}$ for some n, and then use Lemma 5.2.7 below to show that some power of that lift is invariant.) Now we show that some power of y lifts to $HD(n-1)_{**}^{A\,\square_{D(n-1)}\,\mathbf{F}_2}$; since $D(0) = A$, then downward induction on n will finish the proof.

Since y is invariant under the $A\,\square_{D(n)}\,\mathbf{F}_2$-coaction, then it is also invariant under the coaction of $D(n-1)\,\square_{D(n)}\,\mathbf{F}_2$ (since the latter is a quotient Hopf algebra of the former). So y represents a class at the E_2-term of the change-of-rings spectral sequence
$$E_2^{s,t,u}(\mathbf{F}_2) = \operatorname{Ext}_{\mathbf{F}_2[\xi_n^{2^n}]}^{s,u}(\mathbf{F}_2, \operatorname{Ext}_{D(n)}^{t,*}(\mathbf{F}_2, \mathbf{F}_2)) \Rightarrow \operatorname{Ext}_{D(n-1)}^{s+t,u}(\mathbf{F}_2, \mathbf{F}_2).$$
Also, by assumption, y lies in the (t,u)-plane, say $y \in E_2^{0,q,v}$. Choose m large enough so that $m + q - v - 1$ is positive. By Proposition 5.2.2 (with $X = S^0$), we know that for some c and r, we have $E_r^{s,t,u} = 0$ when $ms + t - u > c$ (hence the same is true for $E_{r'}^{s,t,u}$, for all $r' \geq r$).

For each $i \geq 0$, Proposition 1.4.11 tells us that the possible differentials on y^{2^i} are
$$d_{j+1}(y^{2^i}) \in E_{j+1}^{j+1, 2^i q - j, 2^i v},$$
for $j \geq 2^i$. Choose i so that
$$2^i > \max(r - 1, \frac{c - m}{m + q - v - 1}),$$
and fix $j \geq 2^i$. Then we have a vanishing plane at the E_{2^i+1}-term (and hence at the E_{j+1}-term); we claim that the element $d_{j+1}(y^{2^i})$ lies above the vanishing plane, and

so is zero. We just have to verify the inequality specified by the vanishing plane:
$$\begin{aligned} m(j+1) + 2^i q - j - 2^i v &= (m-1)j + m + 2^i(q-v) \\ &\geq (m-1)2^i + m + 2^i(q-v) \\ &= 2^i(m+q-v-1) + m \\ &> \frac{c-m}{m+q-v-1}(m+q-v-1) + m \\ &= c. \end{aligned}$$

Hence y^{2^i} is a permanent cycle. For degree reasons, it cannot be a boundary; hence it gives a nonzero element of E_∞, and hence a nonzero element of $HD(n-1)_{**}$.

It only remains to show that y^{2^i}, or at least some power $y^{2^{i+j}}$, is invariant under the A-coaction. Let $\rho\colon HD(n-1)_{**} \to HD(n)_{**}$ denote the Hurewicz (restriction) map. This map detects nilpotence: if $x \in \ker\rho$, then x is nilpotent. (This follows from Remark 5.1.7(c), for instance; alternatively, this is the main inductive step in proving Theorem 5.1.5.) Hence by Lemma 5.2.7 below, $y^{2^{i+j}}$ is invariant for some j.

This completes the verification of condition (ii) of Definition 5.1.1(b), and hence the proof of Theorem 5.1.2. □

We have used the following in the case when R and S are commutative, so there are no concerns about centrality of particular elements. In the next chapter, we use the lemma in its full generality.

LEMMA 5.2.7. *Suppose that R and S are A-comodule algebras, with an A-linear map $\rho\colon R \to S$ that detects nilpotence: every $x \in \ker\rho$ is nilpotent. Given a central element $z \in R$ so that $\rho(z) \in S$ is invariant under the A-coaction, then z^{2^n} is also invariant, for some n.*

PROOF. Since $\rho(z)$ is invariant, then the coaction on z is of the form
$$z \longmapsto 1 \otimes z + \sum_i a_i \otimes x_i,$$
where each x_i is in $\ker\rho$, and hence is nilpotent. Since z is central, then so is each x_i by Lemma 5.2.8 below. Thus we see that
$$z^{2^n} \longmapsto 1 \otimes z^{2^n} + \sum_i a_i^{2^n} \otimes x_i^{2^n}.$$
This is a finite sum (since these are comodules), so for n sufficiently large, z^{2^n} is invariant. □

LEMMA 5.2.8. *Suppose that R is an A-comodule algebra. Given an element $z \in R$ with A-coaction*
$$z \longmapsto 1 \otimes z + \sum_i a_i \otimes x_i,$$
where the elements a_i are linearly independent, if z is central, then so is each x_i.

PROOF. This is an easy verification. We let $\psi\colon R \to A \otimes R$ be the coaction map on R. We prove that x_i commutes with each element y in R by induction on the number of summands in the coaction on y. If y is invariant—that is, if $\psi(y) = 1 \otimes y$—then the commutator $[y, z]$ maps to $1 \otimes [y, z] + \sum a_i \otimes [y, x_i]$. Since

z is central, then $[y, z] = 0$; since the a_i are linearly independent, then we conclude that $[y, x_i] = 0$ for each i.

Now suppose that the coaction on y has n terms:

$$\psi(y) = 1 \otimes y + \sum_{j=1}^{n-1} b_j \otimes y_j.$$

By coassociativity, each y_j has fewer than n terms in its diagonal, and so commutes with each x_i. So the formula $\psi[y, z] = 0$ gives the same identity as above: $1 \otimes [y, z] + \sum a_i \otimes [y, x_i] = 0$. Again, we conclude that $[y, x_i] = 0$ for each i. □

5.3. Nilpotence and F-isomorphism via quasi-elementary quotients

In this section we use our F-isomorphism and nilpotence theorems for HD—Theorems 5.1.2 and 5.1.5—to prove analogous theorems for the quasi-elementary quotients of A—Theorems 5.1.3 and 5.1.6.

5.3.1. Nilpotence: Proof of Theorem 5.1.6.
We start with Theorem 5.1.6, because we use it in the proof of Theorem 5.1.3. We need a few preliminary results. Suppose we have a map $f \colon S^0 \to X$. We define $X^{\wedge \infty}$ to be the homotopy colimit of the following diagram:

$$S^0 \xrightarrow{f} X \xrightarrow{f \wedge 1} X \wedge X \xrightarrow{f \wedge 1 \wedge 1} X \wedge X \wedge X \to \cdots,$$

and we let $f^{\wedge \infty} \colon S^0 \to X^{\wedge \infty}$ be the obvious map. We recall the following from [**HS98**, Lemma 2.4].

LEMMA 5.3.1. *Given a map $f \colon S^0 \to X$ and a ring spectrum E with unit map $\eta \colon S^0 \to E$, the following are equivalent.*

(a) $E \wedge X^{\wedge \infty} = 0$.
(b) $\eta \wedge f^{\wedge \infty} \colon S^0 \to E \wedge X^{\wedge \infty}$ is zero.
(c) $\eta \wedge f^{\wedge n} \colon S^0 \to E \wedge X^{\wedge n}$ is zero for $n \gg 0$.
(d) $1_E \wedge f^{\wedge n} \colon E = E \wedge S^0 \to E \wedge X^{\wedge n}$ is zero for $n \gg 0$.

For integers $q > r \geq 0$, we define the following quotient Hopf algebras of D:

$$D_r = D/(\xi_1, \ldots, \xi_r),$$
$$D_{r,q} = D_r/(\xi_{r+2}^{2^{r+1}}, \ldots, \xi_q^{2^{r+1}}).$$

See Figure 5.3.A for their profile functions. Recall from Corollary 2.1.8 that the maximal quasi-elementary quotients of A are called $E(r)$ for $r \geq 0$.

LEMMA 5.3.2. *We have $\varinjlim_r HD_r = H\mathbf{F}_2$ and $\varinjlim_q HD_{r,q} = HE(r)$.*

These are homotopy colimits, as usual.

PROOF. We leave this as an exercise. □

LEMMA 5.3.3. *Let i, j, q, r, and R be integers.*

(a) *Suppose that $q > r \geq i \geq 0$ and $q - r \geq j \geq 0$, and consider the Hopf algebra $B = D_{r,q}/(\xi_{r+1}^{2^{i+1}}, \xi_{q+1}^{2^{r+1+j}})$. By Lemma 2.1.2, there are Hopf algebra extensions*

$$E[\xi_{r+1}^{2^i}] \to B \to C_1,$$
$$E[\xi_{q+1}^{2^{r+j}}] \to B \to C_2,$$

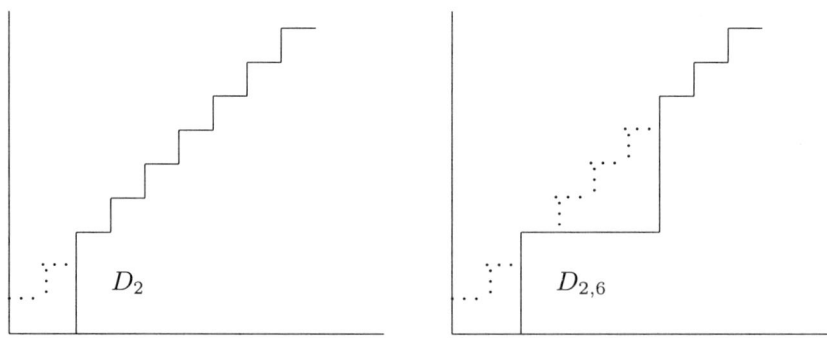

FIGURE 5.3.A. Profile functions for D_r and $D_{r,q}$.

leading to elements $h_{r+1,i}$ and $h_{q+1,r+j}$ in HB_{**}. Then $h_{r+1,i}h_{q+1,r+j}$ is nilpotent in HB_{**}.

(b) Whenever $q > r$ and $r \geq i \geq 0$, there is a Bousfield equivalence
$$\left\langle h_{r+1,i}^{-1} H(D_r/(\xi_{r+1}^{2^{i+1}})) \right\rangle = \left\langle h_{r+1,i}^{-1} H(D_{r,q}/(\xi_{r+1}^{2^{i+1}})) \right\rangle.$$

(c) Whenever $q > R$, we have a Bousfield class decomposition
$$\langle HD \rangle = \langle HD_{R+1} \rangle \vee \bigvee_{r=0}^{R} \bigvee_{i=0}^{r} \left\langle h_{r+1,i}^{-1} H(D_r/(\xi_{r+1}^{2^{i+1}})) \right\rangle$$
$$= \langle HD_{R+1} \rangle \vee \bigvee_{r=0}^{R} \bigvee_{i=0}^{r} \left\langle h_{r+1,i}^{-1} H(D_{r,q}/(\xi_{r+1}^{2^{i+1}})) \right\rangle.$$

PROOF. For (a), the nilpotence of $h_{r+1,i}h_{q+1,r+j}$ is due to Lin [**Lin77a**]; we recall the statement of this result in Theorem B.2.1(b). (The nilpotence of this product is also the content of [**Wil81**, Theorem 6.4], as well as being essentially equivalent to the classification of quasi-elementary quotients of A in Propositions 2.1.7 and 2.1.12.)

Part (b) follows from part (a) and Corollary 1.5.2, by induction: one can get from $D_r/(\xi_{r+1}^{2^{i+1}})$ to $D_{r,q}/(\xi_{r+1}^{2^{i+1}})$ by dividing out by one $\xi_t^{2^s}$ at a time, where $s \geq r+1$. In other words, one has a sequence of extensions of the form
$$E[\xi_t^{2^s}] \to B \to C,$$
and hence cofibrations of the form
$$HB \xrightarrow{h_{ts}} HB \to HC.$$
One inverts $h_{r+1,i}$ in each term; then by the nilpotence of $h_{r+1,i}h_{ts}$, one gets an equivalence of Bousfield classes
$$\langle h_{r+1,i}^{-1} HB \rangle = \langle h_{r+1,i}^{-1} HC \rangle.$$
Part (c) is similar to part (b). □

PROOF OF THEOREM 5.1.6. We imitate the proof of [**HS98**, Theorem 3]. As in that proof, one can reduce to the smash-nilpotence case, and using Spanier-Whitehead duality, one can reduce to the case where $F = S^0$. Suppose we have a map $f\colon S^0 \to X$ so that $1_{HE} \wedge f = 0$ for all quasi-elementary E. We want to show that $1_{HD} \wedge f^{\wedge n}\colon HD \to HD \wedge X^{\wedge n}$ is zero for some n; then Theorem 5.1.5(c) will tell us that f is smash-nilpotent. By Lemma 5.3.1, $1_{HD} \wedge f^{\wedge n}$ is zero if and only if $HD \wedge X^{\wedge \infty} = 0$. We use a Bousfield class argument to show this.

By assumption, for all quasi-elementary E, $1_{HE} \wedge f = 0$, and hence $HE \wedge X^{\wedge \infty} = 0$. By Lemma 5.3.3(c), we must show that
$$HD_{R+1} \wedge X^{\wedge \infty} = 0,$$
for some R, and then that
$$h_{r+1,i}^{-1} H(D_{r,q}/(\xi_{r+1}^{2^{i+1}})) \wedge X^{\wedge \infty} = 0,$$
for all $r \leq R$, $i \leq r$, and some $q > R$.

First we show that $HD_{R+1} \wedge X^{\wedge \infty} = 0$. By Lemma 5.3.1, it is equivalent to show that $\eta \wedge f^{\wedge \infty}\colon S^0 \to HD_{R+1} \wedge X^{\wedge \infty}$ is zero. Let R go to infinity; then by Lemma 5.3.2, the map
$$S^0 \to \varinjlim_R HD_{R+1} \wedge X^{\wedge \infty} = H\mathbf{F}_2 \wedge X^{\wedge \infty}$$
is null. Since homotopy commutes with direct limits, then for some R, the map $S^0 \to HD_{R+1} \wedge X^{\wedge \infty}$ is null.

One uses the same argument to show that
$$h_{r+1,i}^{-1} H(D_{r,q}/(\xi_{r+1}^{2^{i+1}})) \wedge X^{\wedge \infty} = 0,$$
but using the second equality of Lemma 5.3.2, rather than the first. $\square$

5.3.2. F-isomorphism: Proof of Theorem 5.1.3. Now we work on Theorem 5.1.3: the map
$$\pi_{**} S^0 \to \left(\varprojlim_{\mathscr{Q}} HE_{**} \right)^A$$
is an F-isomorphism. First we construct the coaction of A on $\varprojlim_{\mathscr{Q}} HE_{**}$. We let $\overline{\mathscr{Q}}$ denote the full subcategory of $\mathscr{Q}$ consisting of the conormal quasi-elementary quotient Hopf algebras of A. Since the maximal quasi-elementary quotients are conormal, we see that $\overline{\mathscr{Q}}$ is final in $\mathscr{Q}$; hence we have
$$\varprojlim_{\mathscr{Q}} HE_{**} = \varprojlim_{\overline{\mathscr{Q}}} HE_{**}.$$
On the right we have an inverse limit of comodules over A; we give this the induced A-comodule structure. So we have (since taking invariants is an inverse limit)
$$\left(\varprojlim_{\mathscr{Q}} HE_{**} \right)^A = \left(\varprojlim_{\overline{\mathscr{Q}}} HE_{**} \right)^A$$
$$= \varprojlim_{\overline{\mathscr{Q}}} \left(HE_{**}^A \right)$$
$$= \varprojlim_{\overline{\mathscr{Q}}} \left(HE_{**}^{A/\!/E} \right).$$

To finish the proof of Theorem 5.1.3, we show that we can compute HD_{**} up to F-isomorphism in terms of the coefficient rings HE_{**} for E quasi-elementary.

PROPOSITION 5.3.4. *The natural map*
$$HD_{**} \to \varprojlim_{\mathcal{Q}} HE_{**},$$
is an F-isomorphism, as is the induced map
$$HD_{**}^A \to \left(\varprojlim_{\mathcal{Q}} HE_{**}\right)^A.$$

See Proposition 6.5.1 for an explicit computation of the ring $\varprojlim_{\mathcal{Q}} HE_{**}$.

PROOF OF THEOREM 5.1.3. This follows immediately from Theorem 5.1.2 and Proposition 5.3.4. □

PROOF OF PROPOSITION 5.3.4. We need to show that the restriction map
$$\rho \colon HD_{**} \to \varprojlim_{E \in \mathcal{Q}} HE_{**}$$
is an F-isomorphism, so we need to show two things: every element in the kernel of f is nilpotent, and we can lift some 2^nth power of any element in the codomain of f.

By the nilpotence theorem 5.1.6 (or more precisely, by the generalization in Remark 5.1.7(c)), we know that an element $z \in HD_{**}$ is nilpotent if and only if $z^n \in \ker \rho$ for some n.

The second statement follows from the stronger claim that for every quasi-elementary quotient E, some 2^nth power of any element in HE_{**} lifts to HD_{**}. This in turn follows, almost directly, from a result of Hopkins and Smith (stated as Theorem 0.4.10), combined with the fact that the Hopf algebra D is a direct limit of finite-dimensional Hopf algebras. We explain.

As above, we write D_r for the quotient Hopf algebra $D_r = D/(\xi_1, \ldots, \xi_r)$ of D. Then each D_r is conormal in D, the kernel $K_r = D \square_{D_r} \mathbf{F}_2$ is finite-dimensional, and we have $D = \varinjlim_r K_r$. Note that this colimit stabilizes in any given degree.

Given a quasi-elementary quotient E of D, we define the quotient F_r of E to be the pushout of $E \leftarrow D \to D_r$, and we let E_r be the Hopf algebra kernel of $E \to F_r$. In other words, we have the following diagram of Hopf algebra extensions:

$$\begin{array}{ccccc} K_r & \longrightarrow & D & \longrightarrow & D_r \\ \downarrow & & \downarrow & & \downarrow \\ E_r & \longrightarrow & E & \longrightarrow & F_r. \end{array}$$

(In other words, the profile function for F_r is the intersection of the profile functions of D_r and E—cf. the Hopf algebra $B(n)$ in Proposition 2.1.5.)

Now, fix an element $y \in HE_{**}$. Since $E = \varinjlim_r E_r$, then for r sufficiently large, y is in the image of the inflation map $\mathrm{Ext}_{E_r}^{**}(\mathbf{F}_2, \mathbf{F}_2) \to \mathrm{Ext}_E^{**}(\mathbf{F}_2, \mathbf{F}_2)$, say $\tilde{y} \mapsto y$. Also, $K_r \to E_r$ is a quotient map of finite-dimensional graded connected commutative Hopf algebras; let res_{K_r, E_r} denote the induced map on Ext. Theorem 0.4.10 says that given $\tilde{y} \in \mathrm{Ext}_{E_r}^{**}(\mathbf{F}_2, \mathbf{F}_2)$, for some n we have $\tilde{y}^{p^n} \in \mathrm{im}(\mathrm{res}_{K_r, E_r})$. Consider the following commutative diagram:

$$\begin{array}{ccc} \mathrm{Ext}^{**}_{K_r}(\mathbf{F}_2,\mathbf{F}_2) & \longrightarrow & \mathrm{Ext}^{**}_{D}(\mathbf{F}_2,\mathbf{F}_2) \\ \downarrow & & \downarrow \\ \mathrm{Ext}^{**}_{E_r}(\mathbf{F}_2,\mathbf{F}_2) & \longrightarrow & \mathrm{Ext}^{**}_{E}(\mathbf{F}_2,\mathbf{F}_2). \end{array}$$

Given $y \in HE_{**} = \mathrm{Ext}^{**}_E(\mathbf{F}_2,\mathbf{F}_2)$, we can pull y back to $\tilde{y} \in \mathrm{Ext}^{**}_{E_r}(\mathbf{F}_2,\mathbf{F}_2)$, and for some n we can lift $\tilde{y}^{2^n}$ to $\mathrm{Ext}^{**}_{K_r}(\mathbf{F}_2,\mathbf{F}_2)$. The image of this lift in $\mathrm{Ext}^{**}_D(\mathbf{F}_2,\mathbf{F}_2)$ is a lift of y^{2^n}, as desired. This finishes the proof that $HD_{**} \to \varprojlim_{\overline{\mathscr{Q}}} HE_{**}$ is an F-isomorphism.

Now we want to show that the induced map on the invariants is also an F-isomorphism. As above, we write $\overline{\mathscr{Q}}$ for the category of conormal quasi-elementary quotients of A. We want to show that the natural map

$$HD^{A \square_D \mathbf{F}_2}_{**} \to \varprojlim_{\overline{\mathscr{Q}}}(HE^{A \square_E \mathbf{F}_2}_{**})$$

is an F-isomorphism. We treat the $A \square_D \mathbf{F}_2$-coaction on HD_{**} as an A-coaction with trivial D-coaction (and similarly for $A \square_E \mathbf{F}_2$ coacting on HE_{**}), so we want to compare HD^A_{**} with $\varprojlim HE^A_{**}$. We write f and $\tilde{f}$ for the maps

$$f \colon HD_{**} \to \varprojlim_{\overline{\mathscr{Q}}} HE_{**},$$

$$\tilde{f} \colon HD^A_{**} \to \varprojlim_{\overline{\mathscr{Q}}} HE^A_{**}.$$

The maximal quasi-elementary quotients are conormal; hence the conormal quasi-elementary quotients are final in the inverse system of all quasi-elementary quotients; hence f is an F-isomorphism. So it is clear that if $x \in HD^A_{**}$ is in the kernel of $\tilde{f}$, then x is nilpotent. Furthermore, given $y \in \varprojlim HE^A_{**}$, we know that y is in the image of f; hence by Lemma 5.2.7, some 2^nth power of y must be in the image of $\tilde{f}$. $\square$

5.4. Further discussion: nilpotence at odd primes

There are two main obstructions to proving the above results at odd primes. The first is clear: we do not have a classification of the quasi-elementary quotients of A; hence we cannot prove the quasi-elementary versions of the theorems. There are two aspects to this: to prove a version of Proposition 5.3.4, one would like to know that the quasi-elementary quotients are in fact quotients of D; and to prove a version of Lemma 5.3.3, one would like to have an odd primary analogue of Theorem B.2.1(b). The second main obstruction is perhaps more important, since it affects the HD versions of the theorems, and more surprising, because the corresponding result at the prime 2 seems rather standard. This is the property that allows one to prove Lemma 5.2.6 and hence Proposition 5.2.2: at the prime 2, the element $h_{nj} = [\xi_n^{2^j}] \in HD(n)_{**}$ is nilpotent, if $j \geq n$ (see Theorem B.2.1(a)).

Recall from Notation 1.2.14 and Lemma 1.2.15 that a primitive y in a Hopf algebra B gives rise to a class $[y] \in \mathrm{Ext}^{1,|y|}_B(k,k)$. When the characteristic p of the ground field k is odd, an even-dimensional primitive y in B gives rise to a class $\beta \widetilde{\mathscr{P}}^0[y]$ in $\mathrm{Ext}^{2,p|y|}_B(k,k)$.

CONJECTURE 5.4.1. *Fix an odd prime p. Fix a quotient Hopf algebra B of A, and suppose that $\xi_t^{p^s}$ is primitive in B. If $s \geq t$, then $b_{ts} = \beta \widetilde{\mathscr{P}^0}[\xi_t^{p^s}]$ is nilpotent in HB_{**}.*

As at the prime 2, we define D as follows:
$$D = A/(\xi_1^p, \xi_2^{p^2}, \xi_3^{p^3}, \dots).$$
One can see from the proofs when $p = 2$ that the odd prime versions of Theorems 5.1.2 and 5.1.5 would follow from this conjecture. (One has to make a slight change in Lemma 5.2.4—as in the proof of Lemma 1.2.15, there are two related short exact sequences of comodules, leading to a cofibration in which the connecting homomorphism is a map of homological degree 2. Other than that, everything goes through as written.)

We discuss the status of Conjecture 5.4.1 in Appendix B.

5.5. Further discussion: miscellany

The nilpotence theorem [**DHS88**] in stable homotopy theory has far-reaching implications for the global structure of the stable homotopy category—see [**Hop87**] and [**Rav92**], for instance. We develop analogues of some of the structural results in the next chapter, but there are still many gaps. We discuss those in Sections 6.7 and 6.9 below.

If Γ is a Hopf algebra, then the quasi-elementary quotients of Γ give the right homology functors to consider for detecting nilpotence, essentially by the definition of "quasi-elementary." Indeed, if Γ is commutative and finite-dimensional, it is not too hard to prove an analogue of Theorem 5.1.6, either directly (as in [**Pal97**]) or via Chouinard's theorem (as in [**HPS97**, Theorem 9.6.10, Remark 9.6.11]). If, in addition, Γ is connected, then Theorem 0.4.10 allows one to prove an analogue of Theorem 5.1.3—see [**Pal97**]. Actually, by Theorem 0.4.10, some power of every element in the inverse limit is invariant, so one gets an F-isomorphism
$$\operatorname{Ext}_\Gamma^{**}(k,k) \to \varprojlim_{\mathscr{Q}} \operatorname{Ext}_E^{**}(k,k),$$
where the inverse limit is over $\mathscr{Q}$, the category of quasi-elementary quotients of Γ.

Here are some related questions: one already has Quillen stratification for group algebras; can one prove it, as we do for A, with vanishing lines or planes? Can one use our approach to Quillen stratification for the Steenrod algebra to study the cohomology of other Hopf algebras? Given a Hopf algebra Γ, one should study the quotient Hopf algebra which is the pullback of the quasi-elementary quotients of Γ, so one might want to assume that those quotients are somewhat well-behaved. Even more generally, in an arbitrary stable homotopy category one could consider a collection of appropriately chosen ring spectra—versions of the spectra HE—and look at the map from S^0 to an appropriate inverse limit of them.

Can one refine the description of $\pi_{**}S^0$ as given by Theorem 5.1.2? In particular, can one describe the kernel, determine the nilpotence height of elements in the kernel, or say which powers of classes in the codomain of
$$\varphi \colon \pi_{**}S^0 \to HD_{**}^{A \square_D \mathbf{F}_2}$$
lift to the domain?

Since one has a version of Quillen stratification for the Steenrod algebra, one might look for analogues of other group-theoretic results. Chouinard's theorem

[**Cho76**] is an example: a kG-module M is projective if and only if it is projective upon restriction to kE for every elementary abelian subgroup E of G. One analogue in Stable(A) might be: for any spectrum X, $X = 0$ if and only if $HE \wedge X = 0$ for every quasi-elementary quotient E of A. This is false, though: as noted in the proof of Theorem 6.6.1, $X = IHD$ is a counterexample, where I is the Brown-Comenetz duality functor. It is not clear what a better analogue would be, though; other statements, such as "for any spectrum X, X is in $\text{loc}(H\mathbf{F}_p)$ if and only if $HE \wedge X$ is in $\text{loc}(H\mathbf{F}_p)$ for every quasi-elementary quotient E of A," seem equally dubious.

CHAPTER 6

Periodicity and other applications of the nilpotence theorems

The nilpotence theorem in ordinary stable homotopy theory [**DHS88**] has a number of important consequences: the periodicity theorem and the thick subcategory theorem of [**HS98**] are examples. In this chapter we study applications of our nilpotence and Quillen stratification theorems—Theorems 5.1.2, 5.1.3, 5.1.5, and 5.1.6.

One of the main results of this chapter is a version of the periodicity theorem: if X is a finite spectrum, then we produce a number of central non-nilpotent elements of the ring $[X, X]_{**}$ via the "ideal of X," which is essentially the kernel of

$$[S^0, HD]_{**} \xrightarrow{-\wedge X} [X, HD \wedge X]_{**}.$$

Equivalently, this gives families of central non-nilpotent elements in the homotopy of any finite ring spectrum. This is our analogue of the periodicity theorem of Hopkins and Smith. We state this precisely in Section 6.1, and we prove it in Sections 6.2–6.4.

Theorem 5.1.2 says that $\pi_{**}S^0$ is F-isomorphic to the A-invariants in HD_{**}; in Section 6.5 we discuss some examples of invariant elements in HD_{**}.

In Section 6.6, we show that the objects that detect nilpotence—HD and $\bigvee_E HE$—have strictly smaller Bousfield classes than that of the sphere. The role of D, as well as the action of A on D, has led us to a conjectured thick subcategory theorem, which we give in Section 6.7. We end the chapter with two sections of miscellany—one on slope supports of finite spectra, and a short section with a few additional questions and remarks.

As in the previous chapter, we work at the prime 2 unless otherwise stated.

6.1. The periodicity theorem

We start this chapter by giving our version of the periodicity theorem. This is a weak analogue of the Quillen stratification theorem 5.1.2, but with nontrivial coefficients. The following definition was motivated by work in modular group representation theory of Carlson, Avrunin, Scott, and others.

DEFINITION 6.1.1. Given a spectrum X, we define the *ideal of X*, $I(X)$, to be the radical of the ideal

$$\ker\left(HD_{**} \xrightarrow{-\wedge X} [X, HD \wedge X]_{**}\right).$$

Recall from the start of Subsection 5.1.2 that $[X, HD \wedge X]_{**}$ is a ring; one can check that the above map is a ring map.

In Section 6.3, we study the ideal $I(X)$. In particular, we show that when X is finite, the ideal $I(X)$ is invariant (Definition 6.3.1) under the A-coaction;

hence the A-coaction on HD_{**} induces one on $HD_{**}/I(X)$. The following definition generalizes Definition 2.4.2, the definition of u_n-map.

DEFINITION 6.1.2. Given an element $y \in \operatorname{Ext}_D^{**}(\mathbf{F}_2, \mathbf{F}_2) = HD_{**}$ and an object X in $\mathsf{Stable}(A)$, we say that a map $z \colon X \to X$ is a y-map if $1_{HD} \wedge z = y^n \wedge 1$ in $[HD \wedge X, HD \wedge X]_{**}$ for some n.

THEOREM 6.1.3 (Periodicity theorem). *Let X be a finite spectrum. For every $y \in HD_{**}$ which maps to an A-invariant element of $HD_{**}/I(X)$, X has a y-map which is central in the ring $[X, X]_{**}$.*

REMARK 6.1.4. (a) There is an equivalent statement involving elements in the homotopy of a finite ring spectrum R, which we give as Theorem 6.2.3 below.
(b) We conjecture that the ring $(HD_{**}/I(X))^A$ is F-isomorphic to the center of $[X, X]_{**}$. See Section 6.7 for this and related issues.
(c) One of the potential flaws of Theorem 6.1.3 is that computing $I(X)$ is difficult, as is finding invariants in $HD_{**}/I(X)$. If one knows which P_t^s-homology groups of X are nonzero, though, one can find some invariants in $HD_{**}/I(X)$, and hence one can use the theorem to produce some self-maps of X. See Theorem 6.8.3 for details.

6.2. y-maps and their properties

In this section we lay the groundwork for proving Theorem 6.1.3. In particular, we study analogues of v_n-maps [**Hop87**, **HS98**]. In ordinary stable homotopy theory, v_n-maps are defined via the Morava K-theories—these are field spectra, and hence have various convenient properties, such as Künneth isomorphisms. Our analogues are defined via HD, and hence are not quite as easy to work with. Nonetheless, our versions of v_n-maps, called "y-maps," share many of the same properties as v_n-maps in ordinary stable homotopy theory; in particular, we show in this section that the property of having a y-map is generic in our setting.

Fix a quotient Hopf algebra B of A which maps onto D. We work in the category $\mathsf{Stable}(B)$. Recall from Remark 5.1.7(c) that the nilpotence theorem 5.1.5 holds in this category: HD detects nilpotence. Note that if E is a ring spectrum and y is an element of $\pi_{**}E$, then y gives a self map of E, namely the composite

$$E = S^0 \wedge E \xrightarrow{y \wedge 1_E} E \wedge E \xrightarrow{\mu} E.$$

We write y for both the homotopy element and the self-map. We start by expanding Definition 6.1.2 a bit:

DEFINITION 6.2.1. Fix $y \in \operatorname{Ext}_D^{**}(\mathbf{F}_2, \mathbf{F}_2) = HD_{**}$, and let X be an object in $\mathsf{Stable}(B)$. A map $z \colon X \to X$ is a y-map if $1_{HD} \wedge z = y^n \wedge 1 \in [HD \wedge X, HD \wedge X]_{**}$ for some n. Similarly, if R is a ring object in $\mathsf{Stable}(B)$, then an element $\alpha \in \pi_{**}R$ is a y-element if for some n, we have $1_{HD} \wedge \alpha = y^n \wedge 1$ as maps $HD \to HD \wedge R$.

REMARK 6.2.2. (a) Given X and z as in the definition, then the condition $1_{HD} \wedge z = y^n \wedge 1$ is equivalent to the statement that $\eta_{HD} \wedge z = y^n \wedge 1$ in $[X, HD \wedge X]_{**}$. This second condition is a natural one to consider, because the change-of-rings isomorphism (Corollary 1.2.7) tells us that $[X, HD \wedge X]_{**}$ is isomorphic to $[X, X]_{**}^D$, the set of D-linear cochain homotopy classes of maps from X to itself. Similarly, given R and α as above, $1_{HD} \wedge \alpha = y^n \wedge 1$ if and only if $\eta_{HD} \wedge \alpha = y^n \wedge 1$.

(b) Note that the set of y-maps from X to itself is in bijection with the set of y-elements in the ring spectrum $X \wedge DX$, by Spanier-Whitehead duality. Indeed, there is a ring isomorphism

$$[HD \wedge X, HD \wedge X]_{**} \cong [HD, HD \wedge X \wedge DX]_{**}.$$

So the following is equivalent to Theorem 6.1.3.

THEOREM 6.2.3 (Periodicity theorem for ring spectra). *Let R be a finite ring spectrum with unit map $\eta\colon S^0 \to R$. Let $I = \sqrt{\ker HD_{**}\eta}$—this is an invariant ideal in HD_{**}. For every $y \in HD_{**}$ which maps to an A-invariant element in HD_{**}/I, there is a central y-element in $\pi_{**}R$.*

The main tool for proving Theorems 6.1.3 and 6.2.3 is the following. See Definition 1.3.7 for the definition of "thick subcategory."

THEOREM 6.2.4. *Suppose that B is a quotient Hopf algebra of A with $A \twoheadrightarrow B \twoheadrightarrow D$. Fix $y \in HD_{**}$. The full subcategory $\mathscr{C}$ consisting of finite objects of $\mathsf{Stable}(B)$ having a y-map is a thick subcategory of $\mathsf{Stable}(B)$.*

In other words, the property of having a y-map is generic.

The proof is a modification of the proof in [**HS98**, Section 3] that having a v_n-map is a generic property. We devote the rest of this section to the details. We start with a variant of the notion of y-map, and a general lemma.

DEFINITION 6.2.5. Fix $y = HD_{**}$, and let X be an object in $\mathsf{Stable}(B)$. A map $z\colon X \to X$ is a *weak y-map* if $HD_{**}z\colon HD_{**}X \to HD_{**}X$ is multiplication by y^n for some n. Similarly, if R is a ring object in $\mathsf{Stable}(B)$, then an element $\alpha \in \pi_{**}R$ is a *weak y-element* if for some n, $HD_{**}\alpha = y^n$ as maps $HD_{**} \to HD_{**}R$.

We should justify this terminology.

LEMMA 6.2.6. *If a self-map $z\colon X \to X$ of an object X in $\mathsf{Stable}(B)$ is a y-map, then it is a weak y-map. If an element $\alpha \in \pi_{**}R$ in the homotopy of a ring spectrum in $\mathsf{Stable}(B)$ is a y-element, then it is a weak y-element.*

PROOF. This is clear. □

Note, though, that since HD_{**} is not likely to satisfy a Künneth isomorphism, then Spanier-Whitehead duality may *not* yield a bijection between weak y-maps of X and weak y-elements in $\pi_{**}(X \wedge DX)$. Having such a bijection is important in the manipulations below, which is why we work with y-maps rather than weak y-maps. Compare this to the situation in ordinary stable homotopy theory, where v_n-maps are defined in terms of Morava K-theories, which do satisfy a Künneth isomorphism; hence v_n-maps are defined in terms of their effect on $K(n)_*$.

We move on to the proof of Theorem 6.2.4. For the remainder of the section, we consider a fixed element y of HD_{**}. We work at the prime 2.

LEMMA 6.2.7. *Let R be a ring spectrum in $\mathsf{Stable}(B)$, and fix weak y-elements α and β in $\pi_{**}R$. If α and β commute, then there exist positive integers i and j so that $\alpha^i = \beta^j$.*

PROOF. We may assume that $HD_{**}(\alpha - \beta) = 0$ by raising α and β to suitable powers; then by the nilpotence theorem 5.1.5, $\alpha - \beta$ is nilpotent. Since α and β commute, then $\alpha^{2^n} = \beta^{2^n}$ for some n. □

LEMMA 6.2.8. *Let R be a finite ring object in $\mathsf{Stable}(B)$, and fix a weak y-element $\alpha \in \pi_{**}R$. For some $i > 0$, the element α^i is central in $\pi_{**}R$.*

PROOF. Let $\ell(\alpha)_*, r(\alpha)_* \in \operatorname{End}(\pi_{**}R)$ denote left and right multiplication by α, respectively. More precisely, $\ell(\alpha)$ is the following self-map of R:

$$R \xrightarrow{\alpha \wedge 1} R \wedge R \xrightarrow{\mu} R,$$

and similarly for $r(\alpha)$. Since $HD_{**}\alpha$ is central in $HD_{**}R$, then $\ell(\alpha) - r(\alpha)$ maps to zero in $\operatorname{End}(HD_{**}R)$; since R is finite, then the nilpotence theorem for self-maps 5.1.5(b) applies to show that $\ell(\alpha) - r(\alpha)$ is nilpotent. Since right multiplication by α commutes with left multiplication by α, we conclude that $\ell(\alpha)^{2^n} = r(\alpha)^{2^n}$ for some n. Since $\ell(\alpha)^{2^n} = \ell(\alpha^{2^n})$ and $r(\alpha)^{2^n} = r(\alpha^{2^n})$, we conclude that α^{2^n} is central in $\pi_{**}R$. □

COROLLARY 6.2.9. *Let R be a finite ring spectrum in* Stable(B). *For any weak y-elements $\alpha, \beta \in \pi_{**}R$, there exist positive integers i and j so that $\alpha^i = \beta^j$.*

Now we use the adjointness between y-maps and y-elements, so we do not use the weak forms any more.

COROLLARY 6.2.10. *Let X be a finite spectrum in* Stable(B), *and let f and g be two y-maps of X. Then:*

(a) *Some power of f is central in $[X, X]$.*
(b) *$f^i = g^j$ for some positive integers i and j.*

COROLLARY 6.2.11. *Suppose that X_1 and X_2 have y-maps y_1 and y_2, respectively. Then there are positive integers i and j so that for every Z and every $h : Z \wedge X_1 \to X_2$, the following diagram commutes:*

$$\begin{array}{ccc} Z \wedge X_1 & \xrightarrow{h} & X_2 \\ {\scriptstyle 1 \wedge y_1^i} \downarrow & & \downarrow {\scriptstyle y_2^j} \\ Z \wedge X_1 & \xrightarrow{h} & X_2 \end{array}$$

PROOF. $DX_1 \wedge X_2$ has two y-maps: $Dy_1 \wedge 1$ and $1 \wedge y_2$. Now we apply Corollary 6.2.10 and Spanier-Whitehead duality. □

We are ready to prove the main theorem of this section, that the property of having a y-map is generic.

PROOF OF THEOREM 6.2.4. If a finite spectrum Y has a (central) y-map f and if X is a retract of Y, then the induced self-map of X

$$X \to Y \xrightarrow{f} Y \to X$$

is easily seen to be a y-map.

Suppose that $X_1 \to X_2 \to X_3$ is a cofibration of finite spectra, and that X_1 and X_2 have y-maps y_1 and y_2, respectively. By replacing y_1 and y_2 by powers, Corollary 6.2.11 tells us that we may assume that the left square in this diagram commutes:

$$\begin{array}{ccccc} X_1 & \longrightarrow & X_2 & \longrightarrow & X_3 \\ {\scriptstyle y_1} \downarrow & & {\scriptstyle y_2} \downarrow & & \downarrow {\scriptstyle y_3} \\ X_1 & \longrightarrow & X_2 & \longrightarrow & X_3. \end{array}$$

Hence there is a map $y_3 \colon X_3 \to X_3$ so that the right square commutes. We claim that y_3^2 is a y-map. Since y_1 and y_2 are y-maps, then $1_{HD} \wedge y_1$ and $1_{HD} \wedge y_2$ are both multiplication by some power of y, say y^n. We compare $1_{HD} \wedge y_3$ and $y^n \wedge 1$ in $[HD \wedge X_3, HD \wedge X_3]$ via the following commutative diagram:

$$\begin{array}{ccccc} HD \wedge X_2 & \longrightarrow & HD \wedge X_3 & \longrightarrow & HD \wedge X_1 \\ {\scriptstyle 1 \wedge y_2 - y^n \wedge 1}\downarrow & & {\scriptstyle 1 \wedge y_3 - y^n \wedge 1}\downarrow & & {\scriptstyle 1 \wedge y_1 - y^n \wedge 1}\downarrow \\ HD \wedge X_2 & \longrightarrow & HD \wedge X_3 & \longrightarrow & HD \wedge X_1. \end{array}$$

(As usual, we are omitting all suspensions from the notation.) The left- and right-hand vertical maps are null by assumption, so a simple diagram chase shows that $(1 \wedge y_3 - y^n \wedge 1)^2$ is null as well. Since we are working in characteristic 2, and since $1 \wedge y_3$ and $y^n \wedge 1$ commute in $[HD \wedge X, HD \wedge X]$, we find that $1 \wedge y_3^2 = y^{2n} \wedge 1$. Hence y_3^2 is a y-map. □

6.3. Properties of ideals

In this section, we establish a key property of the ideal $I(X)$, namely invariance. As usual in this chapter, we work at the prime 2.

Recall from Corollary 1.2.7 that if B is a conormal quotient Hopf algebra of A, then for any objects X and Y of Stable(A), $[X, HB \wedge Y]_{**}$ is the set of B-linear homotopy classes of maps from X to Y, which we write as $[X, Y]_{**}^B$. We define $I_B(X)$ to be the radical of the kernel of

$$[S^0, S^0]_{**}^B \xrightarrow{-\wedge X} [X, X]_{**}^B.$$

Note that this definition makes sense for any object X in Stable(B). When X is defined over A rather than B, then we show in Proposition 6.3.2 that $I_B(X)$ is invariant in the following sense.

DEFINITION 6.3.1. Suppose that R is a commutative comodule algebra over A (i.e., a commutative algebra and a left A-comodule, compatibly); we write ψ for the comodule structure map. We say that an ideal I of R is *invariant* under the A-coaction if for all $x \in I$, $\psi(x)$ lies in $A \otimes I$.

PROPOSITION 6.3.2. *Let B be a conormal quotient Hopf algebra of A.*

(a) *Let X be a finite object of* Stable(B). *Then*

$$\sqrt{\ker(HB_{**} \to [X,X]_{**}^B)} = \sqrt{\mathrm{ann}_{HB_{**}}(HB_{**}X)}.$$

(b) *Let X be a finite object of* Stable(A). *Then $I_B(X)$ is invariant under the A-coaction on HB_{**}.*

The proof of part (a) is based on similar work in [**Ben91b**, Section 5.7].

PROOF. Part (a): Let

$$I = \sqrt{\ker(HB_{**} \to [X, HB \wedge X]_{**})},$$
$$J = \sqrt{\mathrm{ann}_{HB_{**}}(HB_{**}X)}.$$

The Yoneda action of HB_{**} on $HB_{**}X$ factors through the composition action, since HB_{**} is commutative. Hence $I \leq J$. On the other hand, since X is finite, then it is in thick(S^0). So any element of HB_{**} that annihilates $[S^0, HB \wedge X]_{**}$ will also annihilate $[X, HB \wedge X]_{**}$. Hence $J \leq I$.

Part (b): Now we show that $I(X)$ is invariant under the A-coaction. By Remark 1.2.13, both HB_{**} and $HB_{**}X$ are $A\square_B\mathbf{F}_p$-comodules, and one can check that the action map
$$HB_{**} \otimes HB_{**}X \to HB_{**}X$$
is a map of $A\square_B\mathbf{F}_p$-comodules. Suppose that $y \in I(X)$, and that under the $A\square_B\mathbf{F}_p$-coaction, y maps to $\sum_i a_i \otimes y_i$. (We may assume that the a_i's are linearly independent elements of A.) We want to show that each y_i is in $I(X)$; i.e., that some power of y_i annihilates $HB_{**}X$.

Fix $x \in HB_{**}X$. If we assume that $y^{2^k}x = 0$, then we claim that $y_i^{2^k}x = 0$ for each i. Suppose that under the $A\square_B\mathbf{F}_p$-coaction, we have

$$(6.3.3) \qquad x \longmapsto 1 \otimes x + \sum_{j=1}^n b_j \otimes x_j.$$

We prove, by induction on n, that $y_i^{2^k}x = 0$ for all i. When $n = 0$ (i.e., when x is primitive in $HB_{**}X$), then we have
$$HB_{**}X \to (A\square_B\mathbf{F}_p) \otimes HB_{**}X,$$
$$0 = y^{2^k}x \longmapsto (\sum_i a_i \otimes y_i)^{2^k}(1 \otimes x) = \sum_i a_i^{2^k} \otimes y_i^{2^k}x.$$

Since the a_i's are linearly independent, we conclude that $y_i^{2^k}x = 0$ for all i.

Suppose that $y_i^{2^k}$ annihilates every element of HB_{**} which has at most n terms in its diagonal, and fix x with diagonal as in (6.3.3). Then by coassociativity, each x_j has at most n terms in its diagonal, so we have
$$0 = y^{2^k}x \longmapsto (\sum_i a_i \otimes y_i)^{2^k}(1 \otimes x + \sum_j b_j \otimes x_j)$$
$$= \sum_i (a_i^{2^k} \otimes y_i^{2^k}x + \sum_j a_i^{2^k}b_j \otimes y_i^{2^k}x_j)$$
$$= \sum_i a_i^{2^k} \otimes y_i^{2^k}x.$$

Hence $y_i^{2^k}x = 0$ for each i. $\square$

6.4. The proof of the periodicity theorem

In this section we prove Theorem 6.1.3. Fix a finite object X in $\mathsf{Stable}(A)$. If $y \in HD_{**}$ maps to an invariant element in $HD_{**}/I(X)$, then we want to show that X has a y-map. The basic pattern of the proof is the same as that of Theorem 5.1.2: we inductively work our way from D to A via the Hopf algebras $D(n)$ (defined in Section 5.2).

Since $D(n)$ is a conormal quotient of A for each n, then $[X, HD(n) \wedge X]_{**} = HD(n)_{**}(X \wedge DX)$ is an A-comodule, by Remark 1.2.13. As noted there, the $D(n)$-coaction is trivial, so there is an induced coaction of $A\square_{D(n)}\mathbf{F}_2$; we also have a coaction of the quotient $D(n-1)\square_{D(n)}\mathbf{F}_2$. The statement we want to prove by downward induction on n is: X has a y-map when viewed as an object in $\mathsf{Stable}(D(n))$—i.e., there is a y-map in $[X, HD(n) \wedge X]_{**}$—and that y-map is invariant under the A-coaction (and hence under the $D(n-1)\square_{D(n)}\mathbf{F}_2$-coaction).

By Lemma 5.2.1, HD_{**} is the colimit of the algebras $HD(n)_{**}$, so $y \in HD_{**}$ lifts to $HD(n)_{**}$ for all sufficiently large n; since X is finite, then for some n, the element $y \wedge 1_X \in [X, HD \wedge X]_{**}$ lifts to a y-map $\tilde{y} \in [X, HD(n) \wedge X]_{**} = [X, X]_{**}^{D(n)}$. To start the induction, we need to check that some power of $\tilde{y}$ is invariant under the A-coaction. Let $I_{D(n)}(X)$ be the radical of the kernel of

$$HD(n)_{**} \xrightarrow{-\wedge X} [X, HD(n) \wedge X]_{**}.$$

By Proposition 6.3.2, $I_{D(n)}(X)$ is invariant under the A-coaction. We have a commutative diagram of A-comodules:

$$\begin{array}{ccc} HD(n)_{**}/I_{D(n)}(X) & \longrightarrow & HD_{**}/I(X) \\ {\scriptstyle -\wedge X}\downarrow & & \downarrow{\scriptstyle -\wedge X} \\ [X, HD(n) \wedge X]_{**} & \longrightarrow & [X, HD \wedge X]_{**}. \end{array}$$

By choice of n, the element $y \in HD_{**}$ lifts to $y' \in HD(n)_{**}$; since y maps to an invariant in $HD_{**}/I(X)$, then Lemma 5.2.7 says that some 2^kth power of y' maps to an invariant as well. Since the diagram is one of A-comodules, the image of $(y')^{2^k}$ in $[X, HD(n) \wedge X]_{**}$ is invariant under the A-coaction, as desired. This starts the induction.

(To apply Lemma 5.2.7, we need to know that the map $HD(n)/I_{D(n)}(X) \to HD_{**}/I(X)$ detects nilpotence, but this follows from the fact that $HD(n)_{**} \to HD_{**}$ detects nilpotence.)

Now, assume that X has a y-map when viewed as an object in $\mathsf{Stable}(D(n))$. We want to show that X still has a y-map, but when viewed as an object in $\mathsf{Stable}(D(n-1))$; in other words, we want to show that under the map

$$[X, HD(n-1) \wedge X]_{**} \to [X, HD(n) \wedge X]_{**},$$

some power of the y-map in the target lifts to the source. In $\mathsf{Stable}(D(n-1))$, for any object Y we have the Adams spectral sequence based on $HD(n)$:

$$\operatorname{Ext}^{s,u}_{D(n-1) \square_{D(n)} \mathbf{F}_2}(\mathbf{F}_2, HD(n)_{t,*}(Y)) \Rightarrow [S^0, Y]^{D(n-1)}_{s+t,u}.$$

(As in Section 5.2, we are writing this Adams spectral sequence using the change-of-rings grading of Proposition 1.4.9.) In particular, if we let W_j be defined as in Lemmas 5.2.3 and 5.2.4, then we have

(6.4.1) $\qquad \operatorname{Ext}^{s,u}_{D(n-1) \square_{D(n)} \mathbf{F}_2}(\mathbf{F}_2, HD(n)_{t,*}(X \wedge W_j \wedge DX \wedge DW_j))$

$$\Rightarrow [S^0, X \wedge W_j \wedge DX \wedge DW_j]^{D(n-1)}_{**}$$

$$= [X \wedge W_j, X \wedge W_j]^{D(n-1)}_{**}.$$

Lemma 5.2.6 tells us that $\operatorname{thick}(X) = \operatorname{thick}(W_j \wedge X)$ for any $j \geq n$. Hence by Theorem 6.2.4, it suffices to show that $X \wedge W_j$ has a y-map for some j.

The idea is that since W_j approximates the difference between $D(n)$ and $D(n-1)$, then for j sufficiently large, the endomorphisms of $X \wedge W_j$ over $D(n-1)$ should be more or less the same as the endomorphisms of X over $D(n)$. Since X has a y-map over $D(n)$, then $X \wedge W_j$ should have one over $D(n-1)$.

The details are as follows. Since X is finite, then $X \wedge DX$ is (i_0, j_0)-connective for some i_0 and j_0. The comodule W_j is nonzero only between degrees 0 and $2^j |\xi_n|$, so $X \wedge DX \wedge DW_j$ is $(i_0, j_0 - 2^j|\xi_n|)$-connective. By Lemma 5.2.3, we see that the E_2-term for the Adams spectral sequence (6.4.1) has the following vanishing plane:

$$E_2^{s,t,u}(X \wedge W_j \wedge DX \wedge DW_j) = 0 \text{ when } 2^j|\xi_n|s + t - u + j_0 - 2^j|\xi_n| > 0.$$

The inductive hypothesis tells us that we have an A-invariant y-map in $[X, HD(n) \wedge X]_{**}$. We use $\tilde{y}$ to denote this element, as well as its image in $[X \wedge W_j, HD(n) \wedge X \wedge W_j]_{**}^{D(n-1)}$. Since the map

$$[X, HD(n) \wedge X]_{**} \xrightarrow{-\wedge W_j} [X \wedge W_j, HD(n) \wedge X \wedge W_j]_{**}$$

is a map of $D(n-1) \square_{D(n)} \mathbf{F}_2$-comodules, then the image of $\tilde{y}$ in the target is also invariant. Hence $\tilde{y}$ represents an element at the E_2-term of the spectral sequence. We claim that, when j is large enough, $\tilde{y}$ is a permanent cycle.

Suppose that $\tilde{y} \in \text{Ext}_{D(n)}^{p,q}(X, X)$; then it gives a class in $E_2^{0,p,q}$. The rth differential on this class would lie in $E_r^{r,p-r+1,q}$. So we only have to check that for some j, this group is above the vanishing plane for all $r \geq 2$, and hence zero. We check our inequality:

$$2^j|\xi_n|r + (p - r + 1) - q + j_0 - 2^j|\xi_n| \stackrel{?}{>} 0.$$

Whatever p, q, and j_0 are, we can choose j large enough so that this holds for all $r \geq 2$. Hence $\tilde{y}$ is a permanent cycle in the spectral sequence for $X \wedge W_j$ for $j \gg 0$; it obviously cannot support a differential, so it survives to give a nonzero class at E_∞. Since the resulting self-map of X over $D(n-1)$ restricts to the y-map $\tilde{y}$ over $D(n)$, then it is a y-map over $D(n-1)$. Now we consider the map

$$[X, HD(n-1) \wedge X]_{**} \to [X, HD(n) \wedge X]_{**}.$$

This is a map of A-comodule algebras which detects nilpotence. By assumption, the element $\tilde{y}$ in the codomain is invariant under the A-coaction, and we know that its lift is a y-map, and hence some power of it is central. Lemma 5.2.7 says that some further power is invariant under the A-coaction, as desired. This completes the inductive step, and with it, the proof of Theorem 6.1.3.

REMARK 6.4.2. We have used the nilpotence theorem part of the "Quillen stratification" theorem 5.1.2 (i.e., we have used Theorem 5.1.5) in the proof of Theorems 6.2.4 and 6.1.3. We have not used the other part of Theorem 5.1.2—that some power of any invariant element in HD_{**} lifts to $\pi_{**}S^0$. This in fact follows from the periodicity theorem, so it gives us an alternate proof of that part of Theorem 5.1.2.

6.5. Computation of some invariants in HD_{**}

Theorem 5.1.3 gives an F-isomorphism

$$\pi_{**}S^0 \to (\varprojlim_{\mathscr{D}} HE_{**})^A.$$

We compute $\varprojlim HE_{**}$ in Proposition 6.5.1 below; in Proposition 6.5.6 we give a formula for the coaction of A on this inverse limit; and then we give a few examples of invariant elements. The computations in this section first appeared in [**Pal99**].

At the prime 2, the maximal quasi-elementary quotient Hopf algebras of A are the $E(m)$, $m \geq 0$. We recall from Corollary 2.1.8 and Proposition 2.1.9 their

definition and the computation of their coefficient rings:
$$E(m) = A/(\xi_1, \ldots, \xi_m, \xi_{m+1}^{2^{m+1}}, \xi_{m+2}^{2^{m+1}}, \xi_{m+3}^{2^{m+1}}, \ldots),$$
$$HE(m)_{**} = \mathbf{F}_2[h_{ts} \mid t \geq m+1, 0 \leq s \leq m].$$

The bidegrees of the polynomial generators are given by $|h_{ts}| = (1, |\xi_t^{2^s}|)$. Since it is easy to see the effects of the maps in $\mathcal{Q}$ on the coefficients, we immediately have the following.

PROPOSITION 6.5.1. *There is an algebra isomorphism*
$$\varprojlim HE_{**} \xrightarrow{\cong} \mathbf{F}_2[h_{ts} \mid 0 \leq s < t] / (h_{ts}h_{vu} \mid u \geq t),$$
where $|h_{ts}| = (1, |\xi_t^{2^s}|)$. *Hence there is an F-isomorphism*
$$HD_{**} \xrightarrow{F\text{-iso}} \mathbf{F}_2[h_{ts} \mid 0 \leq s < t] / (h_{ts}h_{vu} \mid u \geq t).$$

The second sentence follows from Proposition 5.3.4, which says that there is an F-isomorphism $HD_{**} \to \varprojlim HE_{**}$.

We want to describe the coaction of A on $\varprojlim HE_{**}$. The coaction of $A \,\square_{E(m)} \mathbf{F}_2$ on $HE(m)_{**}$ is determined by the coaction on the polynomial generators h_{ts}. Furthermore, since the coaction preserves homological degree—$HE(m)_{i*}$ is an A-comodule for each i—then the coaction on each h_{ts} only involves individual generators h_{ij}, not monomials.

PROPOSITION 6.5.2. *Fix $m \geq 0$. Let $\chi\colon A \to A$ denote the conjugation map of A, and for $n \geq 1$ let $\zeta_n = \chi(\xi_n)$. Let $\xi_0 = \zeta_0 = 1$. Under the coaction map*
$$HE(m)_{**} \to (A \,\square_{E(m)} \mathbf{F}_2) \otimes HE(m)_{**},$$
we have
$$h_{ts} \longmapsto \sum_{j=0}^{m-s} \sum_{i=m+1}^{t-j} \zeta_j^{2^s} \xi_{t-i-j}^{2^{i+j+s}} \otimes h_{i,j+s}.$$

The ζ_j part comes, essentially, from the right coaction of A on itself, while the ξ_{t-i-j} comes from the left coaction. See Lemma 6.5.5 for a few useful facts about χ and the ζ_n's.

PROOF. (We assume that $t > m+1$ and that $s < m$; the special cases when $t = m+1$ or $s = m$ are even easier to deal with.) First we find all of the terms in the coaction on h_{ts} of the form $a \otimes h_{ij}$ for $a \in A$ primitive, i.e., with $a = \xi_1^{2^n}$ for some n. For degree reasons, the only possible such terms are $\xi_1^{2^s} \otimes h_{t-1,s+1}$ and $\xi_1^{2^{s+t-1}} \otimes h_{t-1,s}$. By computations as in the proofs of Lemmas A.3 and A.5 of [**Pal96b**], we see that both of these terms do appear; that is, we can see that
$$h_{ts} \longmapsto 1 \otimes h_{ts} + \zeta_1^{2^s} \otimes h_{t-1,s+1} + \xi_1^{2^{s+t-1}} \otimes h_{t-1,s} + \text{other terms},$$
where the "other terms" are of the form $b \otimes h_{ij}$, with $b \in A$ non-primitive. (We have also used Lemma 6.5.5(a) to replace $\xi_1^{2^s}$ with $\zeta_1^{2^s}$.) So the formula given in the proposition is "correct on the primitives"; once we have verified that the formula is co-associative, we will have finished the proof. This verification is a straight-forward (although slightly messy) computation; it could be left to the diligent reader, but due to lack of space constraints, we include it in Lemma 6.5.3 below. □

LEMMA 6.5.3. *The formula*

$$h_{ts} \longmapsto \sum_{j=0}^{m-s} \sum_{i=m+1}^{t-j} \zeta_j^{2^s} \xi_{t-i-j}^{2^{i+j+s}} \otimes h_{i,j+s}.$$

defines a coassociative coaction of A on $HE(m)_{**}$.

PROOF. We write Δ for the coproduct on A, and we write ψ for the coaction map of A on $HE(m)_{**}$. We have to verify that $(\Delta \otimes 1) \circ \psi = (1 \otimes \psi) \circ \psi$, so we compute both of these on h_{ts}. Lemma 6.5.5(b) tells us what $\Delta(\zeta_n)$ is; using that, we compute $(\Delta \otimes 1) \circ \psi(h_{ts})$:

$$(\Delta \otimes 1)(\psi(h_{ts})) = (\Delta \otimes 1)\left(\sum_{j=0}^{m-s}\sum_{i=m+1}^{t-j} \zeta_j^{2^s} \xi_{t-i-j}^{2^{i+j+s}} \otimes h_{i,j+s}\right)$$

$$= \sum_{j=0}^{m-s}\sum_{i=m+1}^{t-j} \Delta(\zeta_j^{2^s})\Delta(\xi_{t-i-j}^{2^{i+j+s}}) \otimes h_{i,j+s}$$

$$= \sum_{j=0}^{m-s}\sum_{i=m+1}^{t-j} \left(\sum_{n=0}^{j} \zeta_n^{2^s} \otimes \zeta_{j-n}^{2^{s+n}}\right)\left(\sum_{q=0}^{t-i-j} \xi_{t-i-j-q}^{2^{i+j+s+q}} \otimes \xi_q^{2^{i+j+s}}\right) \otimes h_{i,j+s}$$

(6.5.4)
$$= \sum_{j=0}^{m-s}\sum_{i=m+1}^{t-j}\sum_{n=0}^{j}\sum_{q=0}^{t-i-j} \zeta_n^{2^s} \xi_{t-i-j-q}^{2^{i+j+s+q}} \otimes \zeta_{j-n}^{2^{s+n}} \xi_q^{2^{i+j+s}} \otimes h_{i,j+s}.$$

On the other hand, for $(1 \otimes \psi) \circ \psi(h_{ts})$ we have

$$(1 \otimes \psi)(\psi(h_{ts})) = (1 \otimes \psi)(\sum_{\ell=0}^{m-s}\sum_{k=m+1}^{t-\ell} \zeta_\ell^{2^s} \xi_{t-k-\ell}^{2^{k+\ell+s}} \otimes h_{k,\ell+s})$$

$$= \sum_{\ell=0}^{m-s}\sum_{k=m+1}^{t-\ell} \zeta_\ell^{2^s} \xi_{t-k-\ell}^{2^{k+\ell+s}} \otimes \psi(h_{k,\ell+s})$$

$$= \sum_{\ell=0}^{m-s}\sum_{k=m+1}^{t-\ell} \zeta_\ell^{2^s} \xi_{t-k-\ell}^{2^{k+\ell+s}} \otimes \sum_{b=0}^{m-\ell-s}\sum_{a=m+1}^{k-b} \zeta_b^{2^{\ell+s}} \xi_{k-a-b}^{2^{a+b+\ell+s}} \otimes h_{a,b+\ell+s}$$

$$= \sum_{\ell=0}^{m-s}\sum_{k=m+1}^{t-\ell}\sum_{b=0}^{m-\ell-s}\sum_{a=m+1}^{k-b} \zeta_\ell^{2^s} \xi_{t-k-\ell}^{2^{k+\ell+s}} \otimes \zeta_b^{2^{\ell+s}} \xi_{k-a-b}^{2^{a+b+\ell+s}} \otimes h_{a,b+\ell+s}.$$

Now we replace a and b with the variables $i = a$ and $j = b + \ell$, and interchange the order of summation:

$$= \sum_{j=0}^{m-s}\sum_{i=m+1}^{t-j}\sum_{\ell=0}^{j}\sum_{k=i+j-\ell}^{t-\ell} \zeta_\ell^{2^s} \xi_{t-k-\ell}^{2^{k+\ell+s}} \otimes \zeta_{j-\ell}^{2^{\ell+s}} \xi_{k+\ell-i-j}^{2^{i+j+s}} \otimes h_{i,j+s}.$$

Finally, we let $n = \ell$ and $q = k - i - j + \ell = k - i - j + n$, to get the formula in (6.5.4). This finishes the proof. □

We have used the following facts about the conjugation map χ. See [**Mil58**, Lemma 10] for a formula for ζ_n in terms of the ξ_j's.

LEMMA 6.5.5. *As above, we write ζ_n for $\chi(\xi_n)$.*

(a) *For any j, $\xi_1^j = \zeta_1^j$.*

(b) *For any n*, $\Delta(\zeta_n) = \sum_{i=0}^{n} \zeta_i \otimes \zeta_{n-i}^{2^i}$, *where $\zeta_0 = 1$*.

PROOF. Part (a): Since $\mathbf{F}_2[\xi_1]$ is a sub-Hopf algebra of A, then one can compute $\chi(\xi_1^j)$ in that sub-Hopf algebra, and the results will hold in A. We conclude that $\xi_1^j = \zeta_1^j$ for any j.

Part (b): The conjugation map χ is an anti-homomorphism—see [**Swe69**, Proposition 4.0.1]; in other words, $\Delta \circ \chi = (\chi \otimes \chi) \circ T \circ \Delta$, where $T : A \otimes A \to A \otimes A$ is the twist map, $T(a \otimes b) = b \otimes a$. So the formula for $\Delta(\xi_n)$ gives the formula for $\Delta(\zeta_n)$. □

Since the $E(m)$'s are the maximal quasi-elementary quotients, the formula in Proposition 6.5.2 for the coaction of A on $HE(m)_{**}$ yields the following corollary.

PROPOSITION 6.5.6. *Under the coaction map*
$$\varprojlim_{\mathscr{Q}} HE_{**} \to A \otimes \varprojlim_{\mathscr{Q}} HE_{**},$$
we have
$$h_{ts} \longmapsto \sum_{j=0}^{\lfloor \frac{s+t-1}{2} \rfloor} \sum_{i=j+s+1}^{t-j} \zeta_j^{2^s} \xi_{t-i-j}^{2^{i+j+s}} \otimes h_{i,j+s}.$$

Milnor's formula for the conjugation map χ in [**Mil58**, Lemma 10] says that $\chi(\xi_n)$ always has a summand of the form $\xi_1^{2^n-1}$, so we can pick out all of the terms of the form $\xi_1^k \otimes h_{i,j}$ to translate this formula into an action of A^* on $\varprojlim HE_{**}$:
$$\mathrm{Sq}^n(h_{ts}) = \begin{cases} h_{t-j,s+j} & \text{if } n = 2^{s+j} - 2^s \text{ and } s+j < t-j, \\ h_{t-j-1,s+j} & \text{if } n = 2^{s+t-1} + 2^{s+j} - 2^s \text{ and } s+j < t-j-1, \\ 0 & \text{otherwise.} \end{cases}$$

We give a graphical depiction of the (co)action in Figure 6.5.A, in which we indicate the coaction by the primitives (i.e., the 2^nth powers of ξ_1).

We end the section with some examples.

EXAMPLE 6.5.7. Let R denote the ring $\varprojlim HE_{**}$.

(a) The element $h_{t,t-1} \in R$ in bidegree $(1, |\xi_t^{2^{t-1}}|) = (1, 2^{t-1}(2^t - 1))$ is an invariant for all $t \geq 1$. Indeed, we know that h_{10} lifts to an element of the same name in $\pi_{1,1} S^0$; also h_{21}^4 lifts to an element in $\pi_{4,24} S^0$ (the element known as g or $\overline{\kappa}$—see [**Zac67**]). We do not know which power of $h_{t,t-1}$ lifts for $t \geq 3$.

(b) We have some families of invariants. $h_{20} \in R$ is not invariant: we have
$$h_{20} \longmapsto 1 \otimes h_{20} + \xi_1^2 \otimes h_{10}.$$
But since $h_{10} h_{21} = 0$ in R, the monomial $h_{20}^i h_{21}^j$ is invariant for all $i \geq 0$ and $j \geq 1$. It turns out that more of these elements lift to $\pi_{**} S^0$ than one might expect from Theorem 5.1.2: the elements in the "Mahowald-Tangora wedge" [**MT68**] are lifts of the elements $h_{20}^i h_{21}^j$ for all $i \geq 0$ and $j \geq 8$. (See [**MPT71**]; Zachariou [**Zac67**] first verified this for elements of the form $h_{20}^{2i} h_{21}^{2(i+j)}$ for $i, j \geq 0$.) These elements are distributed over a wedge between lines of slope $\frac{1}{2}$ and $\frac{1}{5}$ (in the Adams spectral sequence $(t-s, s)$ grading).

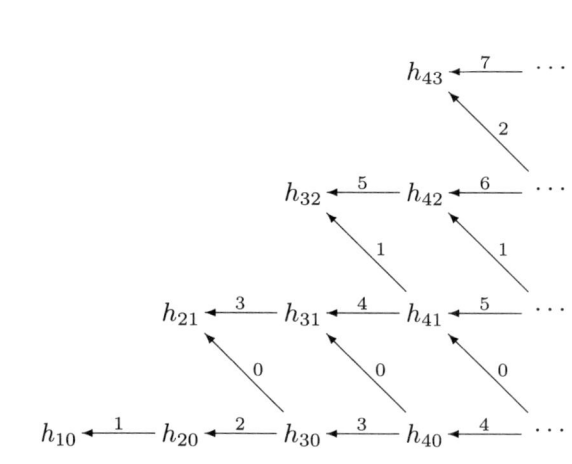

FIGURE 6.5.A. Graphical depiction of coaction of A on $\varprojlim HE_{**}$. An arrow labeled by k represents an action by $\mathrm{Sq}^{2^k} \in A^*$, or equivalently a "coaction" by $\xi_1^{2^k} \in A$—in other words, a term of the form $\xi_1^{2^k} \otimes$ (target) in the coaction on the source.

(c) Similarly, while h_{30} and h_{31} are not invariant, the monomials
$$\{h_{30}^i h_{31}^j h_{32}^k \mid i \geq 0,\ j \geq 0,\ k \geq 1\}$$
are invariant elements. Hence some powers of them lift to $\pi_{**}S^0$. We do not know what powers of them lift, but they will be distributed over a wedge between lines of slope $\frac{1}{6}$ and $\frac{1}{27}$. Continuing in this pattern, we find that for $n \geq 1$, we have sets of invariant elements
$$\{h_{n0}^{i_0} h_{n1}^{i_1} \ldots h_{n,n-1}^{i_{n-1}} \mid i_0, \ldots, i_{n-2} \geq 0,\ i_{n-1} \geq 1\}.$$
The lifts of these elements lie in a wedge between lines of slope $\frac{1}{2^n-2}$ and $\frac{1}{2^{n-1}(2^n-1)-1}$. Hence the family of elements in the Mahowald-Tangora wedge is not a unique phenomenon—we have infinitely many such families, and when $n \geq 3$ they give more than a lattice of points in $\pi_{**}S^0$.

(d) Margolis, Priddy, and Tangora [**MPT71**, p. 46] have found some other non-nilpotent elements, such as $x \in \pi_{10,63}S^0$ and $B_{21} \in \pi_{10,69}S^0$. These both come from the invariant
$$z = h_{40}^2 h_{21}^3 + h_{20}^2 h_{21}^2 h_{41} + h_{30}^2 h_{21} h_{31}^2 + h_{20}^2 h_{31}^3$$
in R. While we do not know which power of z lifts to $\pi_{**}S^0$, we do know that B_{21} maps to the product $h_{20}^3 h_{21}^2 z$, and x maps to $h_{20}^5 z$. In other words, we have at least found some elements in the ideal
$$(z) \subseteq (R)^A$$
which lift to $\pi_{**}S^0$.

(e) Computer calculations have led us to a few other such "sporadic" invariant elements (i.e., invariant elements that do not belong to any family—any

family that we know of, anyway):
$$h_{20}^8 h_{31}^4 + h_{30}^8 h_{21}^4 + h_{21}^{11} h_{31}$$
in bidegree $(12, 80)$, an element in bidegree $(9, 104)$ (a sum of 8 monomials in the variables $h_{i,0}$ and $h_{i,1}$, $2 \leq i \leq 5$), and an element in bidegree $(13, 104)$ (a sum of 12 monomials in the same variables). We do not know what powers of these elements lift to $\pi_{**}S^0$, nor are we aware of any elements in the ideal that they generate which are in the image of the restriction map from $\pi_{**}S^0$.

6.6. Computation of a few Bousfield classes

In [**HPS97**, Section 5.1] we show that if a spectrum E is Bousfield equivalent to the sphere, then E detects nilpotence. This is true in any stable homotopy category, and it is not very deep. While we do not vouch for the depth of the nilpotence theorems of Chapter 5, we at least point out that they are not examples of this generic nilpotence theorem.

Many of the results in this section hold at all primes; some only hold at the prime 2. Unless otherwise indicated, we work at an arbitrary prime p. Bousfield classes are defined in Section 1.5.

THEOREM 6.6.1. *We have $\langle S^0 \rangle > \langle HD \rangle$, and when $p = 2$, $\langle HD \rangle > \bigvee_E \langle HE \rangle$, where the wedge is taken over all quasi-elementary quotients E of A.*

The proof is quite similar to Ravenel's computations in [**Rav84**, Theorem 2.2 and Theorem 3.1] that
$$\langle S^0 \rangle > \langle BP \rangle > \bigvee_{n \geq 0} \langle K(n) \rangle$$
in the ordinary stable homotopy category. We need a few preliminary results, first. Recall that I is the Brown-Comenetz duality functor, defined in Section 1.5. Here is the main lemma.

LEMMA 6.6.2. *Suppose that B and C are quotient Hopf algebras of A that fit into a Hopf algebra extension*
$$B \,\square_C \mathbf{F}_p \hookrightarrow B \twoheadrightarrow C.$$
(a) *Suppose that $\dim_{\mathbf{F}_p} B \,\square_C \mathbf{F}_p = \infty$. If B and C are conormal quotient Hopf algebras of A, then $[HC, HB]_{**} = 0$. Hence $HC \wedge IHB = 0$; hence $\langle HB \rangle > \langle HC \rangle$.*
(b) *Suppose that $\dim_{\mathbf{F}_p} B \,\square_C \mathbf{F}_p < \infty$.*
  (i) *If $B \,\square_C \mathbf{F}_p$ contains some τ_n or some $\xi_t^{p^s}$ with $s < t$, then $\langle HB \rangle > \langle HC \rangle$.*
  (ii) *If not, and if $p = 2$, then $\langle HB \rangle = \langle HC \rangle$.*

PROOF. Part (a): The statement that $[HC, HB]_{**} = 0$ implies the rest of the lemma, by Proposition 1.5.1, so we only have to verify that. By Shapiro's lemma (Example 1.2.4), there is an isomorphism
$$[HC, HB]_{**} \cong \operatorname{Ext}_B^{**}(A \,\square_C \mathbf{F}_p, \mathbf{F}_p).$$
The proof that this Ext group is zero is somewhat technical, so we relegate it to Lemma 6.6.4 below.

Part (b)(i) follows from Corollary 1.5.2 and induction, using the non-nilpotence of the classes h_{ts} (when $s < t$), b_{ts} (when $s < t$), and v_n. Part (b)(ii) is similar, but uses the nilpotence of h_{ts} when $s \geq t$: see Theorem B.2.1. □

We need to complete an Ext calculation to finish the proof of Lemma 6.6.2. To do this, it will be more convenient to work with modules over the dual A^* of A, rather than with A-comodules. This is because we will be using duality, and while the vector space dual of a module is always a module, this is not in general true for comodules without some sort of finiteness hypotheses.

In particular, for any graded Hopf algebra Γ^* over a field k, there is a duality functor
$$d \colon \Gamma^*\text{-Mod} \to \Gamma^*\text{-Mod},$$
defined by setting $dM = \operatorname{Hom}_k(M, k)$. The Γ^*-action on dM is given by $\gamma f(-) = f(\chi(\gamma) \cdot -)$ for $\gamma \in \Gamma^*$ and $f \in \operatorname{Hom}_k(M,k)$, where $\chi \colon \Gamma^* \to \Gamma^*$ is the conjugation map. More generally, there is a Γ^*-action on $\operatorname{Hom}_k(M, N)$ for any Γ^*-modules M and N—see the end of Section 0.3 for details. As noted after Definition 0.2.2, if Γ^* is either commutative or cocommutative, then $\chi^2 = 1$. Hence in this case, if M is of finite type (i.e., if M is finite-dimensional in each degree), then $ddM \cong M$.

We write $\operatorname{Ext}_{\Gamma^*\text{-Mod}}$ for Ext in the category of Γ^*-modules.

LEMMA 6.6.3. *If M and N are Γ^*-modules, then there is an isomorphism*
$$\operatorname{Ext}_{\Gamma^*\text{-Mod}}^{**}(M, N) \cong \operatorname{Ext}_{\Gamma^*\text{-Mod}}^{**}(k, \operatorname{Hom}_k(M, N)).$$

PROOF. We compute $\operatorname{Ext}_{\Gamma^*\text{-Mod}}^{**}(M, N)$ by taking an injective resolution $I_\bullet$ of N and then mapping in M: in other words, we are interested in the cohomology of the cochain complex which is $\operatorname{Hom}_{\Gamma^*\text{-Mod}}(M, I_n)$ in degree n. Of course, Hom_k and $\otimes_k$ are adjoint, so viewing the M in the first entry as $k \otimes M$, we have
$$\operatorname{Hom}_{\Gamma^*\text{-Mod}}(M, I_n) \cong \operatorname{Hom}_{\Gamma^*\text{-Mod}}(k, \operatorname{Hom}_k(M, I_n)).$$
Indeed, the cochain complexes with these terms are isomorphic. The same adjointness shows that $\operatorname{Hom}_k(M, I_n)$ is an injective Γ^*-module, and this expression is an exact functor in M; hence the cochain complex $\operatorname{Hom}_k(M, I_\bullet)$ is an injective resolution of $\operatorname{Hom}_k(M, N)$. □

Here is the missing computation from the proof of Lemma 6.6.2.

LEMMA 6.6.4. *Suppose that B and C are conormal quotient Hopf algebras of A that fit into a Hopf algebra extension*
$$B \square_C \mathbf{F}_p \hookrightarrow B \twoheadrightarrow C.$$
If $\dim_{\mathbf{F}_p} B \square_C \mathbf{F}_p = \infty$, then $\operatorname{Ext}_B^{**}(A \square_C \mathbf{F}_p, \mathbf{F}_p) = 0$.

PROOF. Lemma 0.3.3 tells us that there is an inclusion functor $J \colon A\text{-Comod} \to A^*\text{-Mod}$. Writing $\operatorname{Ext}_{B^*\text{-Mod}}$ for Ext in the category of B^*-modules, we want to show that
$$\operatorname{Ext}_{B^*\text{-Mod}}^{**}(J(A \square_C \mathbf{F}_p), \mathbf{F}_p) = 0.$$
One can see that $J(A \square_C \mathbf{F}_p) = d(A^*//C^*)$, where $A^*//C^* = A^* \otimes_{C^*} \mathbf{F}_p$.

Given an extension of cocommutative Hopf algebras over $\mathbf{F}_p$
$$C^* \hookrightarrow B^* \twoheadrightarrow B^*//C^*,$$

and given a B^*-module M, there is a spectral sequence, the *change-of-rings spectral sequence*, with E_2-term

$$E_2^{***} \cong \mathrm{Ext}^{**}_{B^*//C^*\text{-Mod}}(\mathbf{F}_p, \mathrm{Ext}^{**}_{C^*\text{-Mod}}(\mathbf{F}_p, M)),$$

converging (strongly) to $\mathrm{Ext}^{**}_{B^*\text{-Mod}}(\mathbf{F}_p, M)$. This is the same as the spectral sequence described in Section 1.4 but in the module setting; this module version is more standard and has been described elsewhere, such as [**Sin73**, II, §5] and [**Ben91b**, Section 3.5].

We are interested in $\mathrm{Ext}^{**}_{B^*\text{-Mod}}(d(A^*//C^*), \mathbf{F}_p)$; since $A^*//C^*$ is of finite type, then this Ext group is isomorphic to $\mathrm{Ext}^{**}_{B^*\text{-Mod}}(\mathbf{F}_p, A^*//C^*)$. The change-of-rings spectral sequence for this has E_2-term equal to

$$\mathrm{Ext}^{**}_{B^*//C^*\text{-Mod}}(\mathbf{F}_p, \mathrm{Ext}^{**}_{C^*\text{-Mod}}(\mathbf{F}_p, A^*//C^*)).$$

Since C is a conormal quotient of A, then C^* is a normal sub-Hopf algebra of A^*; hence $A^*//C^*$ is a trivial C^*-module, and the E_2-term is isomorphic to

$$\mathrm{Ext}^{**}_{B^*//C^*\text{-Mod}}(\mathbf{F}_p, A^*//C^* \otimes \mathrm{Ext}^{**}_{C^*\text{-Mod}}(\mathbf{F}_p, \mathbf{F}_p)).$$

Now, $B^*//C^*$ is a sub-Hopf algebra of $A^*//C^*$; hence $A^*//C^*$ is a free module over $B^*//C^*$, by a theorem of Milnor and Moore [**MM65**]. The dual of Lemma 0.3.2 (or equivalently, [**Mar83**, Proposition 12.4]) tells us that $A^*//C^* \otimes \mathrm{Ext}^{**}_{C^*\text{-Mod}}(\mathbf{F}_p, \mathbf{F}_p)$ is also free over $B^*//C^*$; therefore, the E_2-term is isomorphic to

$$\mathrm{Hom}_{B^*//C^*\text{-Mod}}(\mathbf{F}_p, \text{free } B^*//C^*\text{-module}).$$

So it suffices to show that

$$\mathrm{Hom}_{B^*//C^*\text{-Mod}}(\mathbf{F}_p, B^*//C^*) = 0.$$

Dualizing the module $B^*//C^*$ and switching back to the world of comodules, we need to show that

$$\mathrm{Hom}_{B \square_C \mathbf{F}_p}(B \square_C \mathbf{F}_p, \mathbf{F}_p) = 0.$$

This is where the infinite-dimensionality of $B \square_C \mathbf{F}_p$ comes into play. We put the details in Lemma 6.6.5. □

LEMMA 6.6.5. *Suppose B and C are conormal quotient Hopf algebras of A, with $B \twoheadrightarrow C$. If $\dim_{\mathbf{F}_p} B \square_C \mathbf{F}_p = \infty$, then*

$$\mathrm{Hom}_{B \square_C \mathbf{F}_p}(B \square_C \mathbf{F}_p, \mathbf{F}_p) = 0.$$

PROOF. First, for any $B \square_C \mathbf{F}_p$-comodule M and any comodule map $f \colon M \to \mathbf{F}_p$, the kernel of f contains all elements of M which appear in the coproducts of other elements. We claim that when $M = B \square_C \mathbf{F}_p$, the kernel must be all of M, so the map f must be zero.

Fix a nonzero element x of $B \square_C \mathbf{F}_p$; hence x is a polynomial in the ξ_t's (and τ_n's, if p is odd). If we can find an element $\xi_t^{p^s} \in B \square_C \mathbf{F}_p$ which is not a factor of any term of x, then x appears in the coproduct of $\xi_t^{p^s} x$; thus our goal is to find such a $\xi_t^{p^s}$. Since $\dim_{\mathbf{F}_p} B \square_C \mathbf{F}_p = \infty$, the classification of (conormal) quotients of A in Theorem 2.1.1 tells us that there are two cases: either for all $t \gg 0$, there is an s so that $\xi_t^{2^s} \in B \square_C \mathbf{F}_p$, or there is a t so that for all $s \gg 0$, $\xi_t^{p^s} \in B \square_C \mathbf{F}_p$. In either case, we can find an element $\xi_t^{p^s}$ so that $|\xi_t^{p^s}| > |x|$, and this does the job. □

Here are some consequences of Lemma 6.6.2(a).

COROLLARY 6.6.6. $[HD, S^0]_{**} = 0$. *Hence* $HD \wedge IS^0 = 0$; *hence* $\langle S^0 \rangle > \langle HD \rangle$.

PROOF. Apply part (a) of the lemma with $B = A$ and $C = D$. □

COROLLARY 6.6.7. *Let $p = 2$ and let E be a quasi-elementary quotient of A. Then $HE \wedge IHD = 0$; hence $\langle HD \rangle > \langle HE \rangle$.*

PROOF. When E is conormal, Lemma 6.6.2(a) says that $[HE, HD]_{**} = 0$, so the corollary follows in that case. Since the maximal quasi-elementary quotient Hopf algebras of A are conormal quotients of both A and D, then if E is an arbitrary quasi-elementary quotient, there is a maximal quotient E' with $E' \twoheadrightarrow E$. This induces $HE' \to HE$, making HE a module spectrum over HE'. Hence Proposition 1.5.1 tells us that $\langle HE \rangle \le \langle HE' \rangle$. □

COROLLARY 6.6.8. $[A, HD]_{**} = 0$. *Hence if X is a finite object in $\mathrm{loc}(A)$, then $X = 0$.*

PROOF. For the first statement, we apply the lemma with $B = D$ and $C = \mathbf{F}_p$. This then implies that $[Y, HD]_{**} = 0$ for all Y in $\mathrm{loc}(A)$. If X is finite with Spanier-Whitehead dual DX, then $[X, HD]_{**} = HD_{**}(DX)$. So it suffices to show that if X is finite and nontrivial, then $HD_{**}X \ne 0$. Well, $HD_{**}X = 0$ if and only if X is contractible when viewed as a cochain complex of comodules over D, in which case the homology of X must be zero. On the other hand, by the Hurewicz Theorem 1.3.5, if X is finite and nontrivial, then it has nonzero homology. □

PROOF OF THEOREM 6.6.1. It is clear (by Proposition 1.5.1(c), for instance) that
$$\langle S^0 \rangle \ge \langle HD \rangle \ge \bigvee_{\substack{E \text{ quasi-}\\ \text{elem.}}} \langle HE \rangle$$
(where the second inequality is only known to be valid when $p = 2$). By Corollary 6.6.6, the first inequality is strict. When $p = 2$, Corollary 6.6.7 says that the second is, also. □

REMARK 6.6.9. Let $p = 2$ and let $X = \bigvee HE$. One can in fact show a stronger result—that X has no *complement* in HD. Suppose otherwise: suppose that there were an object G so that $\langle HD \rangle = \langle X \rangle \vee \langle G \rangle$ and $0 = X \wedge G$. Smashing the former equality with IHD gives
$$\langle HD \wedge IHD \rangle = \langle X \wedge IHD \rangle \vee \langle G \wedge IHD \rangle,$$
But $HE \wedge IHD = 0$ for all quasi-elementary E by Corollary 6.6.7 and Proposition 1.5.1(b), so we have
$$\langle HD \wedge IHD \rangle = \langle G \wedge IHD \rangle.$$
Also, $\langle IHD \rangle \le \langle H\mathbf{F}_p \rangle \le \langle X \rangle$ by Proposition 1.5.1(a), so we have
$$\langle HD \wedge IHD \rangle = \langle G \wedge IHD \rangle \le \langle G \wedge X \rangle = \langle 0 \rangle.$$
But $HD \wedge IHD \ne 0$.

6.7. Ideals and thick subcategories

In this section, we examine a possible relationship between the ideals $I(X)$ and a classification of thick subcategories of finite objects in $\mathsf{Stable}(A)$. Since this section consists primarily of conjectures, we work for the most part at an arbitrary prime p (although we have more evidence for the conjectures when $p = 2$). There are two subsections: the first one, which is the longer of the two, describes the thick subcategory conjecture and some related results and conjectures. The second discusses rank varieties, and is even more conjectural.

6.7.1. The thick subcategory conjecture.
The Quillen stratification theorem 5.1.2 and the periodicity theorem 6.1.3 provide support for several conjectures about the "global structure" of the category $\mathsf{Stable}(A)$. For instance, we have the following suggested analogue of the result of Hopkins and Smith [**HS98**, Theorem 11], in which they identify the center of $[X, X]_*$ up to F-isomorphism, for any finite p-local spectrum X. Given a ring R, we let $Z(R)$ denote the center of R.

CONJECTURE 6.7.1. *For any finite spectrum X, there is an F-isomorphism*
$$Z[X, X]_{**} \to (HD_{**}/I(X))^A.$$

We should point out there is not even an obvious map between these two rings, but there is a diagram

$$[X, X]_{**}$$
$$\downarrow$$
$$(HD_{**}/I(X))^A \xrightarrow{f} [X, HD \wedge X]_{**}.$$

When $p = 2$, the periodicity theorem implies that large enough pth powers of elements in the image of the horizontal map f lift to the center of $[X, X]_{**}$.

Theorem 6.1.3 also suggests a conjectured classification of thick subcategories of finite spectra in $\mathsf{Stable}(A)$; we spend most of this subsection discussing this conjecture and related ideas. We start with a finite-generation result for the ideal of a finite spectrum. Let R denote the ring

$$R = \varprojlim HE_{**} = \mathbf{F}_2[h_{ts} \mid 0 \leq s < t]/(h_{ts}h_{vu} \mid u \geq t),$$

where $|h_{ts}| = (1, |\xi_t^{2^s}|)$. Proposition 6.5.1 says that there is an F-isomorphism $f \colon HD_{**} \to R$. This gives a bijection between the radical ideals of the two rings HD_{**} and R; given a radical ideal $I \leq HD_{**}$, we write f_*I for the corresponding radical ideal in R.

PROPOSITION 6.7.2. *Let $p = 2$. If X is a finite object of $\mathsf{Stable}(D)$, then $f_*I(X)$ is a finitely generated ideal of R.*

Since we only use this result as motivation for some conjectures, we sketch the proof. (To fill in the details would require a careful discussion of the Atiyah-Hirzebruch spectral sequence in the category $\mathsf{Stable}(A)$.)

SKETCH OF PROOF. To see that $f_*I(X)$ is finitely generated, note that since X is finite, then "most" of D acts trivially on X: as in Subsection 5.3.1, we let D_n denote the following quotient Hopf algebra of D:

$$D_n = D/(\xi_1, \xi_2, \ldots, \xi_n) = \mathbf{F}_2[\xi_{n+1}, \xi_{n+2}, \ldots]/(\xi_{n+1}^{2^{n+1}}, \xi_{n+2}^{2^{n+2}}, \ldots).$$

Then D_n should "act trivially on X" for n large enough. More precisely, there is an Atiyah-Hirzebruch spectral sequence with
$$E_2 = (H\mathbf{F}_p)_{**}X \otimes (HD_n)_{**} \Rightarrow (HD_n)_{**}X.$$
Since X is finite, then its homology $(H\mathbf{F}_p)_{**}X$ is bounded. So if n is large enough, then for degree reasons, the elements in the image of the edge homomorphism
$$(H\mathbf{F}_p)_{**}X \to E_2 = (H\mathbf{F}_p)_{**}X \otimes (HD_n)_{00}$$
are all permanent cycles. This is a spectral sequence of modules over $(HD_n)_{**}$, so everything must be a permanent cycle, and we find that
$$(HD_n)_{**}X \cong (H\mathbf{F}_p)_{**}X \otimes (HD_n)_{**}.$$
Consider the ring map $HD_{**} \to (HD_n)_{**}$. The annihilator in HD_{**} of $HD_{**}X$ is contained in the annihilator in HD_{**} of $(HD_n)_{**}X$. Since
$$\operatorname{ann}_{(HD_n)_{**}}((HD_n)_{**}X) = (0),$$
then $I(X)$ is contained in the radical of the kernel of $HD_{**} \to (HD_n)_{**}$. We can calculate this kernel, up to radical, by imitating the proof of Proposition 5.3.4; we find that we have an F-isomorphism
$$(HD_n)_{**} \to \mathbf{F}_2[h_{ts} \mid s < t, n < t]/(h_{ts}h_{vu} \mid u \geq t),$$
and hence the radical of the kernel of $HD_{**} \to (HD_n)_{**}$ maps to the ideal
$$K = (h_{ts} \mid s < t \leq n)$$
in the Noetherian ring
$$\mathbf{F}_2[h_{ts} \mid s < t \leq n]/(h_{ts}h_{vu} \mid u \geq t) \subseteq R.$$
$\square$

We say that a radical ideal I of HD_{**} is *essentially finitely generated* if the ideal $f_*I \leq \varprojlim HE_{**}$ is finitely generated.

Now we state our main conjecture of this section.

CONJECTURE 6.7.3. *The thick subcategories of finite spectra in $\mathsf{Stable}(A)$ are in one-to-one correspondence with the essentially finitely generated radical ideals of HD_{**} which are invariant under the coaction of $A \square_D \mathbf{F}_p$.*

The conjectured bijection should send an invariant ideal I to the full subcategory $\mathscr{D}(I)$ with objects
$$\{X \text{ finite} \mid I(X) \supseteq I\}.$$
This is clearly a thick subcategory. The other arrow in the bijection should send a thick subcategory $\mathscr{D}$ of the finite objects in $\mathsf{Stable}(A)$ to the ideal $I(\mathscr{D})$, defined by
$$I(\mathscr{D}) = \bigcap_{X \in \mathrm{ob}\ \mathscr{D}} I(X).$$
This is an essentially finitely generated radical invariant ideal by Propositions 6.3.2 and 6.7.2.

EXAMPLE 6.7.4. Here is a little evidence for Conjecture 6.7.3.

(a) By Example 6.5.7(a), we have maps
$$h_{10}\colon S^{1,1} \to S^0,$$
$$h_{21}^4\colon S^{4,24} \to S^0.$$
We let S^0/h_{10} and S^0/h_{21}^4 denote the cofibers of these. One can easily compute the ideals of these cofibers:
$$f_*I(S^0/h_{10}) = \sqrt{(h_{10})},$$
$$f_*I(S^0/h_{21}^4) = \sqrt{(h_{21})}.$$
Since $h_{10}h_{21} = 0$ in $\varprojlim HE_{**}$, then Conjecture 6.7.3 would tell us that
$$\text{thick}(S^0/h_{10}, S^0/h_{21}^4) = \text{thick}(S^0).$$
Let $\mathscr{C} = \text{thick}(S^0/h_{10}, S^0/h_{21}^4)$. Clearly, $\mathscr{C} \subseteq \text{thick}(S^0)$. One can also show that S^0 is contained in $\mathscr{C}$, using the octahedral axiom: the cofiber of the map $h_{10}h_{21}^4\colon S^{5,25} \to S^0$ fits into a cofibration with S^0/h_{10} and S^0/h_{21}^4, and hence is in $\mathscr{C}$. But since $h_{10}h_{21}^4$ is zero in $\pi_{**}S^0$, then this cofiber is just $S^0 \vee S^{4,25}$; hence S^0 is in $\mathscr{C}$.

(b) By Example 6.5.7(b), we have non-nilpotent self-maps of the sphere spectrum called $h_{20}^i h_{21}^j$, for certain exponents i and j. If i and j are both positive, then $f_*I(S^0/(h_{20}^i h_{21}^j)) = \sqrt{(h_{20}h_{21})}$, so Conjecture 6.7.3 would imply that $\text{thick}(S^0/(h_{20}^i h_{21}^j))$ is independent of i and j. Arguing as in part (a), one can see that this is true.

(c) Lastly, we point out that $\text{thick}(S^0) \neq \text{thick}(S^0/(h_{20}^i h_{21}^j))$. For instance, if we let $d_0 = h_{20}^2 h_{21}^2$, then $S^0/d_0 \wedge d_0^{-1}S^0 = 0$, and hence $X \wedge d_0^{-1}S^0 = 0$ for every X in $\text{thick}(S^0/d_0)$. Note that for degree reasons, d_0 induces the zero map on P_t^s-homology for every s and t; hence no P_t^s-homology theory can distinguish between the spectra S^0/d_0 and $S^0 \vee S^0$. Therefore P_t^s-homology groups are not a fine enough invariant to detect thick subcategories. See Section 6.8 for related ideas.

Here is a sketch of part of the proof of Conjecture 6.7.3.

CONJECTURE 6.7.5. *Given any invariant essentially finitely generated radical ideal I of HD_{**}, there is a finite spectrum X so that $I(X) = I$.*

IDEA OF PROOF. Since the ideal $f_*I \leq \varprojlim HE_{**}$ is finitely generated, we can choose generators for it. After raising them to powers, if necessary, we pull those generators back to elements $y_1, y_2, \ldots, y_n \in I$, with $y_k \in HD_{i_k, j_k}$. We order these so that $i_1 \leq i_2 \leq \cdots \leq i_n$; then $(y_1, \ldots, y_k)$ is invariant for each $k \leq n$. For each $k \geq 0$, we want to define spectra X_k inductively so that $I(X_k) = (y_1, \ldots, y_k)$. We start by letting $X_0 = S^0$; then $I(X_0) = \sqrt{(0)}$. Given X_{k-1} with $I(X_{k-1}) = (y_1, \ldots, y_{k-1})$, Theorem 6.1.3 tells us that then X_{k-1} has a y_k-map; we let X_k be the cofiber of this map. Clearly $I(X_k) \supseteq (y_1, \ldots, y_k)$; if we could prove equality, we would be done. □

This would show that the composite $I \mapsto \mathscr{D}(I) \mapsto I(\mathscr{D}(I))$ is the identity. For the composite $\mathscr{D} \mapsto I(\mathscr{D}) \mapsto \mathscr{D}(I(\mathscr{D}))$, one still needs to show that every thick subcategory $\mathscr{D}$ is of the form $\mathscr{D} = \mathscr{D}(I)$ for some invariant ideal I.

In the category $\mathsf{Stable}(kG)$, where G is a finite p-group and k is an algebraically closed field of characteristic p, there are two interesting bijections: one is between thick subcategories of finite spectra and subsets of Proj $\pi_* S^0$ which are closed under specialization, where Proj $\pi_* S^0$ means the set of nonmaximal homogeneous prime ideals in the graded ring $\pi_* S^0$. The other bijection is between Bousfield classes and all subsets of Proj $\pi_* S^0$. (See [**BCR97**] and [**HPb**], as well as Section 6.9, for details.)

By analogy, one might conjecture that there is a bijection between the set of Bousfield classes in Stable(A) and the set of all radical ideals of HD_{**}, but that seems a bit much to expect without further evidence. Indeed, the presence of Bousfield classes $\langle X \rangle$ with $\langle X \rangle < \langle H\mathbf{F}_p \rangle$, such as $\langle X \rangle = \langle IS^0 \rangle$, may be evidence against such a conjecture. Such Bousfield classes make Stable(A) look more similar to the ordinary stable homotopy category than to Stable(kG), despite the Quillen stratification theorem 5.1.3.

We end this subsection with the following result. This is related to Conjecture 6.7.5, and is also a generalization of Theorem 2.5.3, which describes the construction of the spectra $F(u_{d_1}^{j_1}, \ldots, u_{d_m}^{j_m})$.

THEOREM 6.7.6. *Let $p = 2$, and fix an essentially finitely generated invariant radical ideal $I \leq HD_{**}$. Write $I = (y_1, \ldots, y_m)$, and order the generators so that $(y_1, \ldots, y_i)$ is invariant for each $i \leq m$. For any integers $k_1, \ldots, k_m$, there are integers $j_1, \ldots, j_m$ with $k_i \leq j_i$ for each i, so that there is a finite spectrum $F = F(y_1^{j_1}, \ldots, y_m^{j_m})$ satisfying the following.*

(a) *When $I = \sqrt{(0)}$, then $F = S^0$.*
(b) *I is contained in the ideal of F.*
(c) *Hence if $y \in HD_{**}/I$ is invariant, then F has a y-map, u.*
(d) *F is self-dual, as are its y-maps.*
(e) *For any finite spectrum W with $I \leq I(W)$, then W is in thick(F).*
(f) *Hence the Bousfield class of F is independent of the choice of exponents j_i.*
(g) *Fix y and u as in (c). For any finite spectrum W and any y-map v on W, there are integers i and j so that $u^i \wedge 1_W = 1_F \wedge v^j$ as self-maps of $F \wedge W$.*
(h) *Suppose that $\ell_1, \ldots, \ell_m$ are integers so that $F(y_1^{\ell_1}, \ldots, y_m^{\ell_m})$ exists. If $\ell_i \gg j_i$ for each i, then there is a map $F(y_1^{\ell_1}, \ldots, y_m^{\ell_m}) \to F(y_1^{j_1}, \ldots, y_m^{j_m})$ commuting with projection to the top cell.*

See Section 6.9 for an application of this theorem to the construction of certain finite localization functors.

6.7.2. Rank varieties. In this subsection we discuss another body of ideas, based on work of Nakano and the author [**NP98**] (and this was based, in turn, on work of Friedlander and Parshall [**FP86, FP87**], among others). One might want to work over the algebraically closed field $\overline{\mathbf{F}}_p$ instead of $\mathbf{F}_p$ in this subsection.

Let $p = 2$. We let W be the vector space

$$W = \mathrm{Span}_{\mathbf{F}_2}(P_t^s \mid s < t).$$

We view W as being an *inhomogeneous* sub-vector space of the Steenrod algebra A^*, and we let $V_D(\mathbf{F}_2)$ be the following subset of W:

$$V_D(\mathbf{F}_2) = \{y \in W \mid y^2 = 0\}.$$

We do not require that the elements y be homogeneous. One can show (as in the proof of [**NP98**, Theorem 1.7]) that $V_D(\mathbf{F}_2)$ consists precisely of linear combinations of commuting P_t^s's—i.e., it is a union of affine spaces, one such space for each maximal elementary quotient of A. Therefore it is "equal" to the prime ideal spectrum of $\varprojlim_{\mathcal{Q}} HE_{**}$, and hence is homeomorphic to the prime ideal spectrum of HD_{**}. At odd primes, we define

$$W_{\text{ev}} = \text{Span}_{\mathbf{F}_p}(P_t^s \mid s < t),$$
$$W_{\text{odd}} = \text{Span}_{\mathbf{F}_p}(Q_n \mid n \geq 0),$$

and then let $V_D(\mathbf{F}_p)$ be the set of all elements $y = (y_1, y_2)$ in $W_{\text{ev}} \oplus W_{\text{odd}}$ with $y_1^p = 0$ and $y_2^2 = 0$. We do not have as nice a description of $V_D(\mathbf{F}_p)$ at odd primes; because of this, and for other technical reasons, it might be best to restrict the following discussion to the case $p = 2$.

Given an element $y \in V_D(\mathbf{F}_p)$, we can construct the *y-homology spectrum* $H(y)$, just as we did the P_t^s-homology spectrum in Definition 2.2.1. Since y need not be homogeneous, $H(y)$ has only one natural grading, the homological grading. Given a spectrum X, we define its *rank variety*, $V_D(X)$, to be the following subset of $V_D(\mathbf{F}_p)$:

$$V_D(X) = \{y \in V_D(\mathbf{F}_p) \mid H(y)_* X \neq 0\}.$$

We say that a subset V of $V_D(\mathbf{F}_p)$ is *realizable* if $V = V_D(X)$ for some finite spectrum X.

CONJECTURE 6.7.7. *Let X be a finite spectrum. The rank variety $V_D(X)$ determines the thick subcategory generated by X. In other words, there is a bijection between the invariant essentially finitely generated radical ideals of HD_{**} and the realizable subsets of $V_D(\mathbf{F}_p)$.*

This bijection should come about by the following: if X is a finite spectrum, then there should be a homeomorphism (actually, an "inseparable isogeny"—see [**Ben91b**, p. 172]) between the prime ideal spectrum of $HD_{**}/I(X)$ and $V_D(X)$.

6.8. Further discussion: slope supports

In this section, we discuss slope supports. This material was originally introduced in [**Pal96b**] as a P_t^s-homology approach to the periodicity theorem and the thick subcategory conjecture. Since those topics now seem more closely related to $I(X)$, the ideal of X, slope supports appear to be more peripheral. (Because of this, and because the results in this section have appeared elsewhere, we only gives sketches of proofs.) On the other hand, since P_t^s-homology groups do determine vanishing lines and other useful information, studying slope supports may be worthwhile.

The two main results in this section are Proposition 6.8.2, which gives a classification of the possible supports of finite spectra; and Theorem 6.8.3, which uses the periodicity theorem 6.1.3 to produce a number of non-nilpotent self-maps of a finite spectrum.

DEFINITION 6.8.1. Let p be a prime, and let Slopes denote the set of slopes of A (Notation 2.2.8). Given a spectrum X, we define its *slope support* to be the set

$$\text{supp}(X) = \{n \mid Z(n)_{**} X \neq 0\} \subseteq \text{Slopes}.$$

We let Slopes$'$ be the following set:

$$\text{Slopes}' = \begin{cases} \{(t,s) \mid t > s \geq 0\}, & p = 2, \\ \{(t,s) \mid t > s \geq 0\} \cup \{n \mid n \geq 0\}, & p \text{ odd}. \end{cases}$$

Then there is bijection between Slopes and Slopes$'$, with the slope $\frac{p|\xi_t^{p^s}|}{2} \in$ Slopes corresponding to $(t,s) \in$ Slopes$'$, and when p is odd, $|\tau_n| \in$ Slopes corresponding to $n \in$ Slopes$'$. We say that a subset T of Slopes$'$ is *admissible* if T satisfies the following conditions:

(a) When $p = 2$:
$$(t,s) \in T \Rightarrow (t+1, s) \in T,$$
$$(t,s) \in T \Rightarrow (t+1, s-1) \in T,$$
$$\operatorname{card}(\text{Slopes}' \setminus T) < \infty.$$

(b) When p is odd: the above conditions, as well as:
$$n \in T \Rightarrow n+1 \in T.$$

We call such sets "admissible" because they are the possible slope supports of finite spectra, at least at the prime 2.

PROPOSITION 6.8.2. [**Pal96b**, Prop. 3.10 and Thm. A.1] *For any prime p, if $T \subseteq$ Slopes$'$ is admissible, then T is the slope support of some finite spectrum. If $p = 2$, then the converse holds: if X is any finite spectrum, then $\operatorname{supp} X$ is admissible.*

We conjecture that the converse is also true when p is odd. We can at least prove the following: for X finite, if $(Q_n)_{**}X = 0$, then $(Q_{n-1})_{**}X = 0$ (see [**Pal96b**, A.8]). We also note that there are no restrictions on the slope support of an arbitrary spectrum: the connective P_t^s-homology spectrum p_t^s has slope support equal to $\{(s,t)\}$, so any subset of Slopes$'$ may be realized as the support of a wedge of p_t^s's (and similarly at odd primes, using the p_t^s's and the q_n's).

SKETCH OF PROOF. We include a sketch of the proof that every admissible T is the slope support of some finite spectrum, since the corresponding result in [**Pal96b**] assumed that $p = 2$. We focus on the case when p is odd; the reader can imitate this proof or refer to [**Pal96b**] for the $p = 2$ case.

Every admissible $T \subseteq$ Slopes$'$ can be written as
$$T = T(t_1, s_1) \cap \cdots \cap T(t_n, s_n) \cap T(m)$$
for some numbers t_i, s_i, m, where $T(t,s)$ is the largest admissible set not containing (t,s)—i.e., it is the complement of
$$\{(t,s), (t-1,s), (t-1,s+1), (t-2,s), (t-2,s+1), (t-2,s+2), \ldots\}.$$
$T(m)$ is the complement of $\{0, \ldots, m\}$. See Figure 6.8.A. If we can find sufficiently nice finite spectra $X(t,s)$ and $X(m)$ so that $\operatorname{supp}(X(t,s)) = T(t,s)$ and $\operatorname{supp}(X(m)) = T(m)$, then we can realize any slope support T by the spectrum
$$X(t_1, s_1) \wedge \cdots \wedge X(t_n, s_n) \wedge X(m).$$
One might worry that although $Z(d)_{**}X(t_i, s_i) \neq 0$ and $Z(d)_{**}X(t_j, s_j) \neq 0$ for some i, j, and d, one might have $Z(d)_{**}X(t_i, s_i) \wedge X(t_j, s_j) = 0$. This does not happen for injective resolutions of finite comodules—see [**NP98**, Corollary 1.5]—and it certainly does not happen for the examples we construct in the next paragraph. This is what we mean by "sufficiently nice."

We recall from [**Mit85**] (see also Section 4.3) that the quotient Hopf algebra
$$P(n) = A/(\xi_1^{p^{n+1}}, \xi_2^{p^n}, \ldots, \xi_{n+1}^p, \xi_{n+2}, \xi_{n+3}, \ldots; \tau_0, \tau_1, \tau_2, \ldots)$$

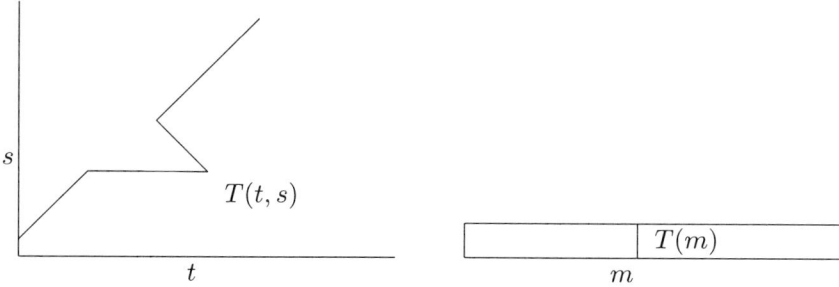

FIGURE 6.8.A. $T(t,s)$ and $T(m)$ as subsets of Slopes$'$.

has the structure of an A-comodule, extending the $P(n)$-comodule structure. The same also holds for
$$V(m) = \Lambda[\tau_0, \ldots, \tau_m].$$
We will use $P(n)$ and $V(m)$ to refer to the Hopf algebras, the comodules, or their injective resolutions, depending on the context. According to Theorem 2.3.6(c), $H(P(n), P_t^s)$ is zero if and only if $\xi_t^{p^s} \neq 0$ in $P(n)$; the corresponding result holds for $V(m)$. Hence as subsets of Slopes$'$, we have
$$\operatorname{supp} P(n) = T(n+1, 0),$$
$$\operatorname{supp} V(m) = T(m).$$

Recall from Section 4.5 that Φ denotes the "pth power" functor on the category of A-comodules: if either $p = 2$ or M is evenly graded, then we define ΦM by
$$(\Phi M)_n = \begin{cases} M_r, & \text{if } n = pr, \\ 0, & \text{if } n \text{ is not divisible by } p, \end{cases}$$
with the comodule structure induced by the Frobenius on A. One can readily see that if $s < t$, then
$$\operatorname{supp} \Phi^s P(t-1) = T(t, s).$$
Hence an injective resolution of the comodule
$$\Phi^{s_1} P(t_1 - 1) \otimes \cdots \otimes \Phi^{s_n} P(t_n - 1) \otimes V(m)$$
has slope support contained in
$$T = T(t_1, s_1) \cap \cdots \cap T(t_n, s_n) \cap T(m).$$
By a Poincaré series argument, one can check that this is exactly the slope support. See [**Pal96b**, A.1] for the proof of the converse when $p = 2$. □

Recall from Notations 2.2.8 and 2.4.1 that for each slope n, there is a non-nilpotent element $u_n \in Z(n)_{**}$. There is a ring map $HD_{**} \to Z(n)_{**}$, and Proposition 6.5.1 implies that some power of u_n lifts to HD_{**}. Hence we can consider the existence of u_n-maps, à la Definition 6.1.2: a u_n-map is a map $z \colon X \to X$ so that z maps to $u_n^j \wedge 1_X$ in $[Z(n), Z(n) \wedge X]_{**}$. Arguing as in Remark 6.2.2(a), one can see that this is the same as the definition of u_n-map given in Definition 2.4.2.

This first appeared for modules as [**Pal96b**, Theorem 4.1]. The proof uses the A-comodule $\frac{1}{2}A(n)$, described in Section 4.3.

THEOREM 6.8.3. *Let $p = 2$. Let X be a finite spectrum and let $T \subseteq$ Slopes be the slope support of X. If $n \in T$ and $T \setminus \{n\}$ is admissible (viewed as a subset of Slopes$'$), then there is a non-nilpotent u_n-map in $[X, X]_{**}$.*

SKETCH OF PROOF. Theorem 6.2.4 tells us that the property of having a u_n-map is generic, so it suffices to find a finite spectrum Y with two properties: Y has a u_n-map, and the thick subcategory generated by Y contains all X with $\text{supp}(X) = T$ as in the statement.

Let $P_t^s = Z(n)$. We claim that $Y = \Phi^s \frac{1}{2} A(t-1)$ has the required properties. First of all, in the terminology used in the proof of Proposition 6.8.2, the slope support of Y is $\text{supp}(Y) = T(t,s) \setminus \{(t,s)\}$. Proposition 4.5.6 describes a map $\Sigma^1 \frac{1}{2} A(t-1) \to \frac{1}{2} A(t-1)$, and one can check that this is an $h_{t,0}$-map; applying Φ^s turns it into an h_{ts}-map.

Theorem 4.5.1 says that the thick subcategory generated by $\frac{1}{2} A(t-1)$ contains all finite spectra Z with $Z(d)_{**}Z = 0$ for all $d < |\xi_t|$; the proof uses vanishing lines to conclude this. In our setting, we replace the vanishing line argument with an application of the nilpotence theorem 5.1.5, to conclude that thick($\Phi^s \frac{1}{2} A(t-1)$) contains all finite spectra with support as described. □

For example, when $X = S^0$, this gives the maps described in Example 6.5.7(a). This result gives incomplete information—it does not give all of the non-nilpotent self-maps of X—but since $Z(n)$-homology groups are easier to compute than the ring $(HD_{**}/I(X))^A$, it is perhaps more accessible than the full periodicity theorem.

6.9. Further discussion: miscellany

We have already mentioned several conjectures related to global structure of the category Stable(A) when $p = 2$—see Section 6.7. We note that some of these issues have been resolved by Hovey and the author for finite-dimensional quotient Hopf algebras of A in [**HPb**] and [**HPa**]. In [**HPb**], we classify the thick subcategories and Bousfield classes in the category Stable(B) for any finite-dimensional quotient B of $A \otimes_{\mathbf{F}_2} \overline{\mathbf{F}}_2$, where $\overline{\mathbf{F}}_2$ is the algebraic closure of $\mathbf{F}_2$. In [**HPa**], we show how to translate a classification of thick subcategories over an algebraically closed field to a similar result over a finite field. Combining the main results of the two papers, we get this result: if B is a finite-dimensional quotient of A, then there is a bijection between thick subcategories of finite objects in Stable(B) and subsets of Proj $\pi_{**}S^0$ which are closed under specialization. Here Proj means the set of nonmaximal bihomogeneous prime ideals, and a set T of primes is closed under specialization if $p \in T$ and $p \subseteq q$ implies $q \in T$.

As far as getting results like Theorem 6.1.3 at odd primes, one has to prove analogues of Theorems 5.1.2 and 5.1.5 first; see Section 5.4 for a discussion of those issues.

We presented some preliminary computations of A-invariants in HD_{**} in Section 6.5; it would be nice to have further results. Extensive computer calculations could be useful, and obviously it would be nice to find new families of invariants. Along similar lines, we could use more information about invariant ideals of HD_{**}— basic properties, examples, and of course a classification would be helpful.

It is also natural to wonder if one can prove a version of the periodicity theorem 6.1.3 which uses quasi-elementary quotients of A instead of D.

6.9. FURTHER DISCUSSION: MISCELLANY

Finally, we point out that one can use the periodicity theorem 6.1.3 to generalize the "killing construction" given in Theorem 3.1.6. Theorem 3.1.6 was proved using the finite spectra $F(u_1^{j_1}, \ldots, u_n^{j_n})$, which were constructed in Theorem 2.5.3 so as to have specific vanishing properties with respect to P_t^s-homology groups. We generalize this by using Theorem 6.7.6 and the finite spectra constructed there. Let I be a finitely generated invariant ideal of HD_{**}, and let $\mathscr{C}$ be the thick subcategory of finite spectra X with $I(X) \supseteq I$. Then we have a finite localization functor L_I^f and a cofibration

$$C_I^f X \xrightarrow{f} X \xrightarrow{g} L_I^f X.$$

If the ideal I is generated by classes u_{d_i}, then $Z(d)_{**}f$ is an isomorphism for $d \notin \{d_i\}$, and $Z(d)_{**}g$ is an isomorphism if $d \in \{d_i\}$. The analogue of Proposition 3.1.3 holds. Our proof of Theorem 3.1.6(b), on the other hand, does not work in this situation; it may be that these finite type and connectivity results only hold when I is as in that theorem, i.e., when I is the radical ideal generated by classes $\{u_d \mid d \leq n\}$, for some n.

APPENDIX A

An underlying model category

Let Γ be a graded commutative Hopf algebra over a field k. In this section we (briefly) describe a model category whose associated homotopy category is equivalent to Stable(Γ). The main results here (Theorems A.1.3 and A.1.4) are due to Hovey; see [**Hov99**] for the details.

Recall that Quillen [**Qui67**] defined the notion of a *closed model category*; this is a category $\mathscr{C}$, like the category of topological spaces, in which one has a notion of a well-behaved homotopy relation between maps. Briefly, a closed model structure is determined by specifying three classes of maps—*weak equivalences*, *fibrations*, and *cofibrations*—satisfying certain properties. See [**Qui67**], as well as [**DS95**] and [**Hov99**], for details. Given this structure, one can define a new category $\mathrm{Ho}\,\mathscr{C}$, the *associated homotopy category*, in which one has inverted the weak equivalences. Rather than work with the homotopy category, it is often more convenient to work with an equivalent category, in which the objects are certain well-behaved objects of $\mathscr{C}$ (the objects which are both "fibrant" and "cofibrant"), and the morphisms are homotopy classes of maps.

These days, by the way, one often says "model category" rather than "closed model category."

Model categories are useful because, while one can do many constructions working entirely in a homotopy category, for certain more delicate operations one needs to work at the "point-set" level—i.e., in the model category. For example, while Adams' definition in [**Ada74**] of the homotopy category of spectra is extremely useful, there are constructions one cannot do unless one has a model category underlying it. Hence today one has various definitions of model categories of spectra, each with its own advantages and disadvantages.

In our work here, we have not needed a model category underlying Stable(Γ). Nonetheless, it seems like a good idea, pedagogically and for future applications, to set one up.

We assume that Γ is a graded commutative Hopf algebra over a field k. We let $\mathsf{Ch}(\Gamma)$ denote the category whose objects are cochain complexes of Γ-comodules (not necessarily injective ones), and whose morphisms are cochain maps. We put a model category structure on $\mathsf{Ch}(\Gamma)$; to do this, we need to specify the weak equivalences, the fibrations, and the cofibrations in $\mathsf{Ch}(\Gamma)$.

NOTATION A.1.1. Given cochain complexes X and Y, let $[X,Y]$ denote the set of cochain homotopy classes of degree zero maps from X to Y. Given a Γ-comodule M, let $S^i M$ denote the cochain complex which is M in degree i, and zero elsewhere. Let $L(k)$ denote an injective resolution of the trivial module k, and let $\mathscr{S}$ denote the set of simple comodules of Γ.

Given an integer i, $M \in \mathscr{S}$, and X any object of $\mathsf{Ch}(\Gamma)$, we define $\pi_i(X; M)$ (the "ith homotopy group of X with coefficients in M") by
$$\pi_i(X; M) = [S^i M, L(k) \otimes X].$$
We say that a map $f\colon X \to Y$ in $\mathsf{Ch}(\Gamma)$ is a *weak equivalence* if and only $\pi_i(f; M)$ is an isomorphism for all integers i and all simple comodules M. We say that a map $f\colon X \to Y$ is a *fibration* if and only if each component $f_n\colon X_n \to Y_n$ of f is an epimorphism with injective kernel. We say that a map $f\colon X \to Y$ is a *cofibration* if and only if each component $f_n\colon X_n \to Y_n$ is a monomorphism.

REMARK A.1.2. One can show (see [**Hov99**]) that a map f is a weak equivalence if and only if $1_{L(k)} \otimes f$ is a cochain homotopy equivalence. Also, note that $\pi_i(X; k) = [S^i k, L(k) \otimes X]$ is the ith cohomology group of the cochain complex of primitives of $L(k) \otimes X$; this is a useful alternate description of $\pi_*(-; k)$.

THEOREM A.1.3. [**Hov99**] *With weak equivalences, fibrations, and cofibrations defined as above, $\mathsf{Ch}(\Gamma)$ is a model category. Its associated homotopy category is equivalent to $\mathsf{Stable}(\Gamma)$.*

Note that with this model structure, every object of $\mathsf{Ch}(\Gamma)$ is cofibrant.

One can in fact give $\mathsf{Ch}(\Gamma)$ the structure of a *cofibrantly generated* model category, as follows; we refer to [**Hov99**] for the proofs. Given a Γ-comodule M, we let $D^n M$ denote the (contractible) cochain complex which is M in degrees n and $n+1$, zero elsewhere, with differential given by the identity map:
$$\cdots \to 0 \to 0 \to M \xrightarrow{1_M} M \to 0 \to 0 \to \cdots.$$
Both S^n (defined in Notation A.1.1) and D^n are functors from $\mathsf{Ch}(\Gamma)$ to itself. We let J denote the following set of maps in $\mathsf{Ch}(\Gamma)$:
$$J = \{D^n f \mid n \in \mathbf{Z},\ f\colon M \hookrightarrow N \text{ an inclusion of finite-dimensional comodules}\}.$$
We define I by
$$I = J \cup \{S^{n+1} M \hookrightarrow D^n M \mid n \in \mathbf{Z},\ M \text{ simple}\},$$
where the map $S^{n+1} M \hookrightarrow D^n M$ is the obvious inclusion.

Here is the second main theorem of this section.

THEOREM A.1.4. [**Hov99**] *With I and J defined as above, $\mathsf{Ch}(\Gamma)$ has the structure of a cofibrantly generated model category, in which I is the set of generating cofibrations and J is the set of generating trivial cofibrations.*

APPENDIX B

Steenrod operations and nilpotence in $\mathrm{Ext}_\Gamma^{**}(k,k)$

Let A be the dual of the mod p Steenrod algebra. In this appendix we recall a few results about Steenrod operations in the cohomology of any commutative Hopf algebra Γ over $\mathbf{F}_p$, and we focus in particular on the case when Γ is a quotient Hopf algebra of A. Then we discuss the nilpotence of certain classes in $\mathrm{Ext}_\Gamma^{**}(\mathbf{F}_p, \mathbf{F}_p)$, for use in proving the nilpotence theorems of Chapter 5.

B.1. Steenrod operations in Hopf algebra cohomology

In this short section we recall a few facts about Steenrod operations in Hopf algebra cohomology. May's paper [**May70**, Section 11] is our basic reference; see also [**Sin73**], [**Wil81**], [**Saw82**], [**BMMS86**], and [**Rav86**] for related results.

Suppose that Γ is a commutative Hopf algebra over the field $\mathbf{F}_p$. Then there are Steenrod operations acting on $\mathrm{Ext}_\Gamma^{**}(\mathbf{F}_p, \mathbf{F}_p)$—as usual, Ext here is taken in the category of Γ-comodules. The operations are of this form:

- If $p = 2$, then for each $n \geq 0$ there is an operation
$$\widetilde{\mathrm{Sq}}^n \colon \mathrm{Ext}_\Gamma^{s,t}(\mathbf{F}_2, \mathbf{F}_2) \to \mathrm{Ext}_\Gamma^{s+n, 2t}(\mathbf{F}_2, \mathbf{F}_2).$$

- If p is odd, then for each $n \geq 0$ there are operations
$$\widetilde{\mathscr{P}}^n \colon \mathrm{Ext}_\Gamma^{s,2t}(\mathbf{F}_p, \mathbf{F}_p) \to \mathrm{Ext}_\Gamma^{s+qn, 2pt}(\mathbf{F}_p, \mathbf{F}_p),$$
$$\beta\widetilde{\mathscr{P}}^n \colon \mathrm{Ext}_\Gamma^{s,2t}(\mathbf{F}_p, \mathbf{F}_p) \to \mathrm{Ext}_\Gamma^{s+qn+1, 2pt}(\mathbf{F}_p, \mathbf{F}_p).$$

Here, $q = 2p - 2$. Note that $\beta\widetilde{\mathscr{P}}^n$ is a single operation, not the composite of two operations.

These satisfy the usual properties of Steenrod operations: the Cartan formula, the Adem relations, and some instability conditions; for example, if $x \in \mathrm{Ext}_\Gamma^{s,t}(\mathbf{F}_p, \mathbf{F}_p)$ (with s and t even if p is odd), then
$$\widetilde{\mathrm{Sq}}^s(x) = x^2, \quad p = 2,$$
$$\widetilde{\mathscr{P}}^{s/2}(x) = x^p, \quad p \text{ odd}.$$

Note that at odd primes, the operations are zero on classes in $\mathrm{Ext}_\Gamma^{s,t}(\mathbf{F}_p, \mathbf{F}_p)$ when t is odd. This is an artifact of the grading conventions on the operations. To remedy this, one can either use a different grading convention (see [**May70**] and [**BMMS86**, IV.2] for two different conventions), or one can define operations indexed by half-integers, as in [**Rav86**, Theorem A1.5.2].

EXAMPLE B.1.1. (a) Fix a commutative Hopf algebra Γ over $\mathbf{F}_p$. Recall from Lemma 0.4.8 that $\mathrm{Ext}_\Gamma^1(\mathbf{F}_p, \mathbf{F}_p)$ is isomorphic to the vector space of primitives in Γ. If x is primitive, then we write $[x]$ for the corresponding element of Ext. We have the following Steenrod operations: when p is odd, then $\widetilde{\mathscr{P}}^0[x] = [x^p]$. At the prime 2, $\widetilde{\mathrm{Sq}}^0[x] = [x^2]$. One can compute $\widetilde{\mathscr{P}}^0$ and $\widetilde{\mathrm{Sq}}^0$ on general Ext classes by a similar formula—these operations are induced by the following map on the cobar complex:

$$[x_1|x_2|\cdots|x_n] \mapsto [x_1^p|x_2^p|\cdots|x_n^p].$$

See [**May70**, Proposition 11.10] for a proof.

(b) Recall that $E[x]$ and $D[x]$ are defined in Notation 0.5.1. If $\Gamma = E[x]$, then $\mathrm{Ext}_\Gamma^{**}(\mathbf{F}_p, \mathbf{F}_p) \cong \mathbf{F}_p[h]$, where $h = [x]$. If $p = 2$, then $\widetilde{\mathrm{Sq}}^1(h) = h^2$, and $\widetilde{\mathrm{Sq}}^n(h) = 0$ when $n \neq 1$. If p is odd (in which case $|x|$ must be odd), then all operations vanish on h (because of our grading conventions).

(c) If p is odd and $\Gamma = D[x]$, then $\mathrm{Ext}_\Gamma^{**}(\mathbf{F}_p, \mathbf{F}_p) \cong \Lambda[h] \otimes \mathbf{F}_p[b]$, where $h = [x]$ and b is the p-fold Massey product of h with itself. In this case, $\beta\widetilde{\mathscr{P}}^0(h) = b$, and all other operations on h are zero; also, $\widetilde{\mathscr{P}}^1(b) = b^p$, and all other operations on b are zero.

B.2. Nilpotence in Ext over quotients of A: $p = 2$

In this section we recall a result of Lin's which we use in the proofs of the main results of Chapter 5. Lin's result first appeared as [**Lin77a**, Theorem 3.1 and Corollary 3.2]. Wilkerson [**Wil81**, Theorem 6.4] has also proved a related result.

Let A be the dual of the mod 2 Steenrod algebra. Recall from Notation 1.2.14 and Remark 2.1.3 that if $\xi_t^{2^s}$ is primitive in a quotient Hopf algebra B of A, then we let $h_{ts} = [\xi_t^{2^s}]$ denote the corresponding element of $HB_{1,*}$.

THEOREM B.2.1. *Suppose that*

$$B = A/(\xi_1^{2^{n_1}}, \xi_2^{2^{n_2}}, \ldots)$$

is a quotient Hopf algebra of A so that for some integer m, the exponents n_i are finite when $i = 1, 2, \ldots, m - 1$. Fix μ so that $\xi_m^{2^\mu}$ is primitive in B.

(a) *If $\mu \geq m$, then $h_{m,\mu}$ is nilpotent in HB_{**}.*

(b) *Fix an integer $\ell \leq m$, and suppose that for some λ, $\xi_\ell^{2^\lambda}$ is primitive in B. If $\ell \leq \mu$, then $h_{\ell,\lambda}h_{m,\mu}$ is nilpotent in HB_{**}.*

Note that if $\mu \geq \max(n_1, \ldots, n_{m-1})$, then $\xi_m^{2^\mu}$ will be primitive. Note also that part (a) is a corollary of part (b)—just set $\ell = m$ and $\lambda = \mu$.

One example of this theorem is that the class $h_{11} = [\xi_1^2] \in \mathrm{Ext}_A^{1,2}(\mathbf{F}_2, \mathbf{F}_2)$ is nilpotent. The proof is easy, and demonstrates the techniques used by Lin to prove the general result: the (reduced) diagonal $\xi_2 \mapsto \xi_1^2 \otimes \xi_1$ in A gives the relation $h_{11}h_{10} = 0$ in $\mathrm{Ext}_A^{**}(\mathbf{F}_2, \mathbf{F}_2)$. Applying the Steenrod operation $\widetilde{\mathrm{Sq}}^1$ gives

$$h_{12}h_{10}^2 + h_{11}^3 = 0,$$

so multiplying through by h_{11} and using $h_{11}h_{10} = 0$ yields $h_{11}^4 = 0$. Note that repeatedly applying $\widetilde{\mathrm{Sq}}^0$ to this relation gives $h_{1n}^4 = 0$ for all $n \geq 1$.

For our purposes, part (a) is one of the key ingredients in the proof that restriction to the quotient Hopf algebra D detects nilpotence. Part (b) is used in the classification of quasi-elementary quotients of A, essentially, and is also used in showing that restricting to the quasi-elementary quotients detects nilpotence.

B.3. Nilpotence in Ext over quotients of A: p odd

In this section we discuss the odd-primary analogue of Theorem B.2.1.

Fix an odd prime p, and let A be the dual of the mod p Steenrod algebra. Recall from Remark 2.1.3 that if $\xi_t^{p^s}$ is primitive in a quotient Hopf algebra B of A, then $h_{ts} = [\xi_t^{p^s}]$ is the corresponding element of $HB_{1,*} = \operatorname{Ext}_B^{1,*}(\mathbf{F}_p, \mathbf{F}_p)$, and $b_{ts} \in HB_{2,*}$ is defined to be $\beta\widetilde{\mathscr{P}}^0(h_{ts})$. Alternatively, b_{ts} is equal to the p-fold Massey product of h_{ts} with itself. If τ_n is primitive in B, we let $v_n = [\tau_n]$ be the corresponding element of $HB_{1,*}$.

For convenience, we restate Conjecture 5.4.1. This would be the odd-primary analogue of Theorem B.2.1(a).

CONJECTURE B.3.1. *Fix integers s and t. Suppose that*
$$B = A/(\xi_1^{p^{n_1}}, \xi_2^{p^{n_2}}, \ldots; \tau_0^{e_0}, \tau_1^{e_1}, \ldots)$$
is a quotient Hopf algebra of A in which $\xi_t^{p^s}$ is nonzero and primitive. If $s \geq t$, then b_{ts} is nilpotent in HB_{**}.

One can also consider a partial analogue of Theorem B.2.1(b): under conditions on s, t, and n, the product $b_{ts}v_n$ is nilpotent.

PROPOSITION B.3.2. *Fix integers s, t, and n. Suppose that B is a quotient Hopf algebra of A in which $\xi_i^{p^s} = 0$ for $i < t$, and $\tau_j = 0$ for $j < n$; hence $\xi_t^{p^s}$ and τ_n are primitive in B. If $n \leq s$, then $b_{ts}v_n$ is nilpotent in $\operatorname{Ext}_B^{**}(\mathbf{F}_p, \mathbf{F}_p)$.*

PROOF. The coproduct
$$\tau_{n+t} \mapsto \sum_{i=0}^{n+t} \xi_{n+t-i}^{p^i} \otimes \tau_i + \tau_{n+t} \otimes 1,$$
together with the conditions on B, gives the following relation in the cobar complex for B:
$$\sum_{i=n}^{n+t-1} h_{n+t-i,i}v_i = 0.$$
Hence $h_{tn}v_n = -\sum_{i=n+1}^{n+t-1} h_{n+t-i,i}v_i$. Applying Steenrod operations gives the following (here one needs to use half-integer indexed operations $\widetilde{\mathscr{P}}^{\frac{n}{2}}$, or one needs to use operations $\mathscr{P}^k$ which are indexed differently):
$$b_{tn}v_n^p = \pm \sum_{i=n+1}^{n+t-1} b_{n+t-i,i}v_i^p,$$
$$b_{t,n+1}v_n^{p^2} = \pm \sum_{i=n+1}^{n+t-1} b_{n+t-i,i+1}v_i^{p^2},$$
$$\vdots$$
$$b_{ts}v_n^{p^{s+n-1}} = \pm \sum_{i=n+1}^{n+t-1} b_{n+t-i,i+s-n}v_i^{p^{s+n-1}}.$$

Since $\xi_i^{p^s} = 0$ for $i < t$, each term on the right hand side of this equation is zero. □

Note that from Miller's computation in [**Mil78**, Corollary 3.6] of the v_1-inverted Ext of the Moore spectrum (see also Subsection 4.6.3), one can see that $b_{10}v_1$ need not be nilpotent; hence this result may be best possible.

Another analogue of Theorem B.2.1(b) would be that under conditions on s, t, u, and v, the product $b_{ts}b_{vu}$ is nilpotent. Conjecture B.3.1 should be a special case of such a result, and since we do not know how to prove this conjecture, we will not be able to prove such an analogue of Theorem B.2.1(b).

For the remainder of this section, we provide evidence for and partial results related to Conjecture B.3.1.

We point out that if $s < t$, then $\mathbf{F}_p[\xi_t]/(\xi_t^{p^{s+1}})$ is a quotient Hopf algebra of B; the cohomology of this quotient is

$$\Lambda[h_{t0}, \ldots, h_{ts}] \otimes \mathbf{F}_p[b_{t0}, \ldots, b_{ts}].$$

The element b_{ts} is non-nilpotent when restricted to this quotient, and hence non-nilpotent in Ext_B. When $s \geq t$, though, surprisingly little seems to be known about the nilpotence (or lack thereof) of b_{ts}. For example, while it is easy to verify that $h_{11}^4 = 0$ in $\mathrm{Ext}_A^{**}(\mathbf{F}_2, \mathbf{F}_2)$ at the prime 2 (see Section B.2), we have not been able to locate or prove a similar result for b_{11} at an arbitrary odd prime. Working at the prime 3, Nakamura [**Nak75**] proved that $b_{11}^2 = h_{11}z$ for some z. Since h_{11} is in odd total degree, then $h_{11}^2 = 0$; hence $b_{11}^4 = 0$.

Note that if an element $\xi_t^{p^s}$ is primitive in a Hopf algebra B, then so is $\xi_t^{p^r}$ for all $r \geq s$.

LEMMA B.3.3. *Fix a Hopf algebra B, and fix integers $s \geq t \geq 1$. Suppose that $\xi_t^{p^s}$ is primitive in B.*

(a) *If b_{ts} is nilpotent, then so is $b_{t,s+1}$.*

(b) *Conversely, assume that $\xi_1^{p^{s-t}} = \cdots = \xi_{t-1}^{p^{s-t}} = 0$ in B, and that $s \geq t$. If $b_{t,s+1}$ is nilpotent, then so is b_{ts}.*

As an application of (a), Nakamura's calculation implies that $b_{1,n}^4 = 0$ for all $n \geq 1$, when $p = 3$. One might conjecture (based on very little evidence) that at any odd prime p, $b_{1,n}^q = 0$ for all $n \geq 1$, where $q = 2(p-1)$. As an application of (b), b_{11} is nilpotent in Ext_B if B is a finite-dimensional quotient of A, because in that case, $\xi_1^{p^k} = 0$ in B for some k, hence $b_{1k} = 0$ in Ext_B.

PROOF. Part (a) follows from the relation $\widetilde{\mathscr{P}}^0(b_{ts}^i) = b_{t,s+1}^i$.

For part (b), we have the following coproduct in B:

$$\xi_{2t} \mapsto \sum_{i=0}^{2t} \xi_{2t-i}^{p^i} \otimes \xi_i.$$

Since $\xi_i^{p^{s-t}} = 0$ when $i < t$, this simplifies to

$$\xi_{2t}^{p^{s-t}} \mapsto \xi_{2t}^{p^{s-t}} \otimes 1 + \xi_t^{p^s} \otimes \xi_t^{p^{s-t}} + 1 \otimes \xi_{2t}^{p^{s-t}}.$$

Also since $\xi_i^{p^{s-t}} = 0$ when $i < t$, then $\xi_t^{p^r}$ is primitive in B for every $r \geq s - t$. So the above coproduct in B translates to the relation $h_{t,s-t}h_{ts} = 0$ in Ext_B. We apply Steenrod operations to this relation: applying $\widetilde{\mathscr{P}}^{p^t} \cdots \widetilde{\mathscr{P}}^p(\beta\widetilde{\mathscr{P}}^1)(\beta\widetilde{\mathscr{P}}^0)$ gives

$$b_{ts}^{p^t+1} = \pm b_{t,s-t}^{p^t} b_{t,s+t}.$$

If $b_{t,s+1}$ is nilpotent, then so is $b_{t,s+t}$, by part (a). Hence so is b_{ts}. □

We have the following conjecture, as a special case of Conjecture B.3.1. To some extent, this would generalize Nakamura's result at the prime 3; however, he determines a smaller nilpotence height (4) than this would (the number $N = p\frac{p^{p-1}-1}{(p-1)^2} = 6$ produced in Subsection B.3.1 below).

CONJECTURE B.3.4. *Fix an odd prime p.*
  (a) *The element $b_{11} \in \operatorname{Ext}_A^{2,2p^2(p-1)}(\mathbf{F}_p, \mathbf{F}_p)$ is nilpotent.*
  (b) *Fix $t \geq 1$ and let $j = \frac{p+1}{2}$. Then b_{tt} is nilpotent in $\operatorname{Ext}_B^{**}(\mathbf{F}_p, \mathbf{F}_p)$, where*
$$B = \mathbf{F}_p[\xi_t, \xi_{2t}, \ldots, \xi_{jt}, \xi_{jt+1}, \xi_{jt+2}, \xi_{jt+3}, \ldots] \otimes \Lambda[\tau_0, \tau_1, \ldots].$$

We have a proof for this conjecture when $p = 3$, so it generalizes Nakamura's result.

PROPOSITION B.3.5. *Conjecture B.3.4 holds when $p = 3$.*

The proof (which contains a gap when $p > 3$), is a bit lengthy, and so we relegate it to a subsection. We also include a few other technical results in that subsection.

B.3.1. Sketch of proof of Conjecture B.3.4, and other results.

SKETCH OF PROOF. The proof involves some Massey product manipulations. May's paper [**May69**] is the standard reference for Massey products; many of the key results are reproduced in [**Rav86**, Section A.1.4].

As remarked after Lemma 1.2.15, the element b_{ts} is the p-fold Massey product of $h_{ts} = [\xi_t^{p^s}]$ with itself.

Part (a): In A, we have the coproduct
$$\Delta \colon \xi_2 \longmapsto \xi_2 \otimes 1 + \xi_1^p \otimes \xi_1 + 1 \otimes \xi_2.$$
This produces the relation $h_{10}h_{11} = 0$ in Ext_A. Applying the Steenrod operation $\beta \widetilde{\mathscr{P}}^0$ gives the relation
$$h_{11}b_{11} - b_0 h_{12} = 0.$$
Then for any $k \geq 1$, we apply $\widetilde{\mathscr{P}}^{p^{k-1}} \cdots \widetilde{\mathscr{P}}^p \widetilde{\mathscr{P}}^1$ to get
$$h_{1,k+1} b_{11}^{p^k} = b_{10}^{p^k} h_{1,k+2}.$$
Using induction gives the following formula, valid for all $k \geq 2$:
(B.3.6) $$h_{11} b_{11}^{1+p+p^2+\cdots+p^{k-2}} = h_{1k} b_{10}^{1+p+p^2+\cdots+p^{k-2}}.$$
So we let
$$N = 1 + 1 + (1+p) + (1+p+p^2) + \cdots + (1+p+p^2+\cdots+p^{p-2})$$
$$= p \frac{p^{p-1}-1}{(p-1)^2}.$$
and we look at b_{11}^N. Recall that b_{11} is the p-fold Massey product $\underbrace{\langle h_{11}, \ldots, h_{11}\rangle}_{p}$.

This Massey product has no indeterminacy.

LEMMA B.3.7. *The element b_{11}^N is contained in the p-fold Massey product*
$$\langle h_{11}, h_{11}b_{11}, h_{11}b_{11}^{1+p}, \ldots, h_{11}b_{11}^{1+p+\cdots+p^{p-2}}\rangle.$$

PROOF. By the definition of Massey products, if $y = \langle a_1, \ldots, a_n \rangle$ (with no indeterminacy), then for any elements $x_1, \ldots, x_n$, we have
$$yx_1 \cdots x_n = \langle a_1, \ldots, a_n \rangle x_1 \cdots x_n$$
$$\in \langle a_1 x_1, \ldots, a_n x_n \rangle.$$
So we apply this to $b_{11}^N = \langle h_{11}, \ldots, h_{11} \rangle b_{11}^{N-1}$. □

By our computations with Steenrod operations (i.e., equation (B.3.6)), we have
$$\langle h_{11}, h_{11} b_{11}, h_{11} b_{11}^{1+p}, \ldots, h_{11} b_{11}^{1+p+\cdots+p^{p-2}} \rangle$$
$$= \langle h_{11}, h_{12} b_{10}, h_{13} b_{10}^{1+p}, \ldots, h_{1p} b_{10}^{1+p+\cdots+p^{p-2}} \rangle.$$

LEMMA B.3.8. *The Massey product*
$$\langle h_{11}, h_{12} b_{10}, h_{13} b_{10}^{1+p}, \ldots, h_{1p} b_{10}^{1+p+\cdots+p^{p-2}} \rangle$$
contains $\langle h_{11}, h_{12}, h_{13}, \ldots, h_{1p} \rangle b_{10}^{N-1} = 0$.

We let $\zeta_n = \chi(\xi_n)$, where $\chi \colon A \to A$ is the canonical anti-automorphism. Hence $\zeta_1 = -\xi_1$, and
$$\Delta(\zeta_n) = \sum_{i=0}^{n} \zeta_i \otimes \zeta_{n-i}^{p^i}.$$

PROOF. We have to prove two things: that the Massey product contains $\langle h_{11}, h_{12}, h_{13}, \ldots, h_{1p} \rangle b_{10}^{N-1}$, and that $\langle h_{11}, h_{12}, h_{13}, \ldots, h_{1p} \rangle = 0$. To prove the first of these, one follows the proof of Lemma B.3.7. To prove the second, we show that for all $n \geq 2$, the n-fold Massey product $\langle h_{10}, h_{11}, h_{12}, \ldots, h_{1,n-1} \rangle = 0$; this goes by induction on n. Once we know this, then applying the Steenrod operation $(\widetilde{\mathscr{P}}^0)^i$ (to either the Massey product or to the proof) gives $\langle h_{1,i}, h_{1,i+1}, h_{1,i+2}, \ldots, h_{1,i+n-1} \rangle = 0$ for any i.

Indeed, we show that for each n, the cobar element $d[\zeta_n]$ is equal to
$$\langle h_{10}, h_{11}, h_{12}, \ldots, h_{1,n-1} \rangle;$$
hence this Massey product is cohomologous to zero. We also show that this Massey product has no indeterminacy. As a consequence, $d[\zeta_n^{p^i}]$ kills $\langle h_{1,i}, \ldots, h_{1,i+n-1} \rangle$.

When $n = 2$, the coproduct
$$\zeta_2 \mapsto \zeta_1 \otimes \zeta_1^p = \xi_1 \otimes \xi_1^p$$
gives have the relation $h_{10} h_{11} = 0$. This starts the induction. When $n = 3$, we have
$$\zeta_3 \longmapsto \zeta_2 \otimes \zeta_1^{p^2} + \zeta_1 \otimes \zeta_2^p.$$
Since $d[\zeta_2] = h_{10} h_{11}$, then $d[\zeta_2^p] = h_{11} h_{12}$, so we have
$$\langle h_{10}, h_{11}, h_{12} \rangle = [\zeta_2] h_{12} + h_{10} [\zeta_2^p] = d[\zeta_3] = 0.$$
Since $h_{1n} \in \text{Ext}_A^{1,p^n q}(\mathbf{F}_p, \mathbf{F}_p)$, then the indeterminacy of this Massey product is of the form
$$h_{10} x + y h_{12},$$
where $x \in \text{Ext}_A^{1,q(p+p^2)}(\mathbf{F}_p, \mathbf{F}_p)$ and $y \in \text{Ext}_A^{1,q(1+p)}(\mathbf{F}_p, \mathbf{F}_p)$. Since we know that $\text{Ext}^{1,*}$ is in one-to-one correspondence with the primitives of A, then $\text{Ext}^{1,i}$ is nonzero only when i is of the form $p^j q$ for some j. Hence both x and y must be zero, and there is no indeterminacy.

Now, we assume that $\langle h_{10}, h_{11}, \ldots, h_{1,i-1}\rangle = 0$ via ζ_i, for all $i < n$. Then $\zeta_i^{p^j}$ kills $\langle h_{1,j}, h_{1,j+1}, \ldots, h_{1,j+i-1}\rangle$ for all $i < n$ and all j, and so

$$\langle h_{10}, h_{11}, \ldots, h_{1,n-1}\rangle = \sum_{i=1}^{n-1} [\zeta_i | \zeta_{n-i}^{p^i}].$$

Taking the coproduct on ζ_n kills this Massey product; there is no indeterminacy for the same reason as when $n = 3$. □

So the Massey product in Lemma B.3.7 contains both b_{11}^N and 0; hence b_{11} is an element of the indeterminacy. Therefore we need some information about that indeterminacy. The indeterminacy of a p-fold Massey product is contained in the union of certain $(p-1)$-fold matric Massey products (see [**May69**, Proposition 2.3]); if every entry in the p-fold Massey product has odd total degree, the same is true of each entry in each matrix in the shorter Massey products.

Here is a general conjecture about "short" Massey products at an odd characteristic. This is the gap in our proof.

CONJECTURE B.3.9. *Fix an odd prime p, and fix $n < p$. Consider an n-fold matric Massey product $\langle V_1, \ldots, V_n\rangle$, in which each entry of each matrix V_i has odd total degree. Then any element of this matric Massey product is nilpotent.*

(As in [**Rav86**, Section A.1.4], whenever we discuss matric Massey products, we assume that the matrices involved have entries with compatible degrees, so that their products have homogeneous degrees, etc. See [**May69**, Notation 1.1] for details.)

The conjecture is trivially true when $n < 3$, by graded commutativity; in particular, it is true when $p = 3$. Indeed, when $p = 3$, we see that every element of the indeterminacy is nilpotent of height p. One can also show that the conjecture is true when taking the Massey product of one-dimensional classes in the cohomology of a space [**Dwy**]; otherwise, we do not have much evidence for it. Meanwhile, it has the following consequence.

CONJECTURE B.3.10. *If $n \leq p$ and if a_i has odd total degree for $i = 1, \ldots, n$, then any element x in the indeterminacy of $\langle a_1, \ldots, a_n\rangle$ is nilpotent.*

This would imply that b_{11}^N is nilpotent, as desired.

Part (b) of the conjecture would be proved similarly. One starts with the relation $h_{t0}h_{tt} = 0$ in Ext_B and applies Steenrod operations to get the following replacement for (B.3.6):

$$h_{tt} b_{tt}^{p^{t-1} + p^{2t-1} + \cdots + p^{kt-1}} = \pm h_{t,(k+1)t} b_{t0}^{p^{t-1} + p^{2t-1} + \cdots + p^{kt-1}}.$$

If we set

$$M = 1 + p^{t-1} + (p^{t-1} + p^{2t-1}) + \cdots + (p^{t-1} + p^{2t-1} + \cdots + p^{(p-1)t-1}),$$

then we get

$$b_{tt}^M = \langle h_{tt}, \ldots, h_{tt} \rangle b_{tt}^{M-1}$$
$$\in \langle h_{tt}, h_{tt} b_{tt}^{p^{t-1}}, h_{tt} b_{tt}^{p^{t-1}+p^{2t-1}}, \ldots, h_{tt} b_{tt}^{p^{t-1}+\cdots+p^{(p-1)t-1}} \rangle$$
$$= \langle h_{tt}, h_{t,2t} b_{t0}^{p^{t-1}}, h_{t,3t} b_{t0}^{p^{t-1}+p^{2t-1}}, \ldots, h_{t,pt} b_{t0}^{p^{t-1}+\cdots+p^{(p-1)t-1}} \rangle$$
$$\ni \langle h_{tt}, h_{t,2t}, h_{t,3t}, \ldots, h_{t,pt} \rangle b_{t0}^{M-1}$$
$$= 0.$$

So Conjecture B.3.10 implies that b_{tt}^M is nilpotent. $\square$

As remarked above, the gap in the proof—Conjecture B.3.9—is not a gap when $p = 3$, so Conjecture B.3.4 holds at the prime 3. Arguing similarly, we have the following.

PROPOSITION B.3.11. *Fix $p = 3$. The element $b_{22} \in \mathrm{Ext}_B^{2,432}(\mathbf{F}_3, \mathbf{F}_3)$ is nilpotent, where $B = A/(\xi_1)$.*

PROOF. The coproduct $\xi_4 \mapsto \xi_2^{p^2} \otimes \xi_2$ gives the relation $h_{20} h_{22} = 0$ in Ext_B. Hence $h_{2,i} h_{2,i+2} = 0$ for all i.

As in the "proof" of Conjecture B.3.4, we find that b_{22}^{16} is contained in the Massey product $\langle h_{22}, h_{24} b_{20}^3, h_{26} b_{20}^{3+9} \rangle$; this Massey product also contains the element $\langle h_{22}, h_{24}, h_{26} \rangle b_{20}^{15}$. We claim that the three-fold Massey product $\langle h_{22}, h_{24}, h_{26} \rangle$ is nilpotent.

This Massey product is defined because the product $h_{2,i} h_{2,i+2}$ is killed by $\xi_4^{3^i}$. So we consider the diagonal on ξ_6:

$$\xi_6 \mapsto \xi_4^{3^2} + \xi_3^{3^3} \otimes \xi_3 + \xi_2^{3^4} \otimes \xi_4$$
$$= \langle h_{20}, h_{22}, h_{24} \rangle + h_{30} h_{33}.$$

We have used the fact that $\xi_3^{3^i}$ is primitive in $A/(\xi_1)$, and hence gives rise to a 1-dimensional Ext class $h_{1,i}$. Hence we have

$$h_{30} h_{33} = -\langle h_{20}, h_{22}, h_{24} \rangle.$$

By graded commutativity, $(h_{30} h_{33})^2 = 0$; hence the same is true of the Massey product.

We find that $b_{22}^{16} = \langle h_{22}, h_{24}, h_{26} \rangle b_{10}^{15} + x$ for some class x in the indeterminacy of a three-fold Massey product. Both x and $\langle h_{22}, h_{24}, h_{26} \rangle$ are nilpotent, and hence so is b_{22}. $\square$

COROLLARY B.3.12. *Fix $p = 3$. The element $b_{22} \in \mathrm{Ext}_C^{2,432}(\mathbf{F}_3, \mathbf{F}_3)$ is nilpotent, where $C = A/(\xi_1^3)$.*

PROOF. If we know that b_{22} is nilpotent in $\mathrm{Ext}_{A/(\xi_1)}$ and we want to know about its nilpotency in $\mathrm{Ext}_{A/(\xi_1^3)}$, then by Lemma 1.2.15 it suffices to determine whether $b_{10} b_{22}$ is nilpotent in $\mathrm{Ext}_{A/(\xi_1^3)}$: b_{22} is nilpotent over C if and only if $b_{10} b_{22}$ is nilpotent over C.

This is easy, though: the coproduct of ξ_3 reduces in C to

$$\xi_3 \mapsto \xi_2^3 \otimes \xi_1,$$

giving the relation $h_{10} h_{21} = 0$ in $\mathrm{Ext}_C^{**}(\mathbf{F}_3, \mathbf{F}_3)$. Applying $\beta \widetilde{\mathscr{P}^0}$ gives

$$h_{11} b_{21} - b_{10} h_{22} = 0,$$

and so $b_{10}h_{22} = 0$ ($h_{11} = 0$ over C, since $\xi_1^3 = 0$ in C). Applying $\beta\widetilde{\mathscr{P}^3}$ then gives
$$b_{10}^3 b_{22} = 0.$$

□

This implies that when $p = 2$, b_{22} is nilpotent in $\operatorname{Ext}_B^{**}(\mathbf{F}_3, \mathbf{F}_3)$ for any quotient Hopf algebra B in which ξ_2^9 is primitive.

Bibliography

[AD73]   D. W. Anderson and D. M. Davis, *A vanishing theorem in homological algebra*, Comment. Math. Helv. **48** (1973), 318–327.

[Ada56] J. F. Adams, *On the cobar construction*, Proc. Nat. Acad. Sci. U.S.A. **42** (1956), 409–412.

[Ada74] J. F. Adams, *Stable homotopy and generalised homology*, University of Chicago Press, Chicago, Ill.-London, 1974, Chicago Lectures in Mathematics.

[AM71]   J. F. Adams and H. R. Margolis, *Modules over the Steenrod algebra*, Topology **10** (1971), 271–282.

[AM74]   J. F. Adams and H. R. Margolis, *Sub-Hopf algebras of the Steenrod algebra*, Math. Proc. Cambridge Philos. Soc. **76** (1974), 45–52.

[AS82]   G. S. Avrunin and L. L. Scott, *Quillen stratification for modules*, Invent. Math. **66** (1982), no. 2, 277–286.

[BC76]   E. H. Brown and M. Comenetz, *Pontryagin duality for generalized homology and cohomology theories*, Amer. J. Math. **98** (1976), 1–27.

[BC87]   D. J. Benson and J. F. Carlson, *Complexity and multiple complexes*, Math. Z. **195** (1987), no. 2, 221–238.

[BCR96] D. J. Benson, J. F. Carlson, and J. Rickard, *Complexity and varieties for infinitely generated modules II*, Math. Proc. Cambridge Philos. Soc. **120** (1996), no. 4, 597–615.

[BCR97] D. J. Benson, J. F. Carlson, and J. Rickard, *Thick subcategories of the stable module category*, Fund. Math. **153** (1997), no. 1, 59–80.

[Ben91a] D. J. Benson, *Representations and cohomology I*, Cambridge University Press, Cambridge, 1991, Basic representation theory of finite groups and associative algebras.

[Ben91b] D. J. Benson, *Representations and cohomology II*, Cambridge University Press, Cambridge, 1991, Cohomology of groups and modules.

[BMMS86] R. R. Bruner, J. P. May, J. E. McClure, and M. Steinberger, H_∞ *ring spectra and their applications*, Lecture Notes in Mathematics, vol. 1176, Springer-Verlag, Berlin-New York, Berlin, 1986.

[Boa]   J. M. Boardman, *Operations on the Adams spectral sequences for Brown-Peterson homology and cohomology*, preprint (1988).

[Bou79a] A. K. Bousfield, *The Boolean algebra of spectra*, Comment. Math. Helv. **54** (1979), no. 3, 368–377.

[Bou79b] A. K. Bousfield, *The localization of spectra with respect to homology*, Topology **18** (1979), no. 4, 257–281.

[CE56]   H. Cartan and S. Eilenberg, *Homological algebra*, Princeton University Press, Princeton, N. J., 1956.

[Cho76] L. G. Chouinard, *Projectivity and relative projectivity over group rings*, J. Pure Appl. Algebra **7** (1976), no. 3, 287–302.

[DHS88] E. S. Devinatz, M. J. Hopkins, and J. H. Smith, *Nilpotence and stable homotopy theory I*, Ann. of Math. (2) **128** (1988), no. 2, 207–241.

[DM88]   D. M. Davis and M. E. Mahowald, v_1-*periodic Ext over the Steenrod algebra*, Trans. Amer. Math. Soc. **309** (1988), no. 2, 503–516.

[DP]   W. G. Dwyer and J. H. Palmieri, *Ohkawa's theorem: There is a set of Bousfield classes*, Proc. Amer. Math. Soc., to appear.

[DS95]   W. G. Dwyer and J. Spaliński, *Homotopy theories and model categories*, Handbook of algebraic topology (Ioan M. James, ed.), North-Holland, Amsterdam, 1995, pp. 73–126.

[Dwy]   W. G. Dwyer, private communication.

[Eis87] D. K. Eisen, *Localized Ext groups over the Steenrod algebra*, Ph.D. thesis, Princeton University, 1987.

[ES96] L. Evens and S. Siegel, *Generalized Benson-Carlson duality*, J. Algebra **179** (1996), no. 3, 775–792.

[FP86] E. M. Friedlander and B. J. Parshall, *Support varieties for restricted Lie algebras*, Invent. Math. **86** (1986), no. 3, 553–562.

[FP87] E. M. Friedlander and B. J. Parshall, *Geometry of p-unipotent Lie algebras*, J. Algebra **109** (1987), no. 1, 25–45.

[FS97] E. M. Friedlander and A. Suslin, *Cohomology of finite group schemes over a field*, Invent. Math. **127** (1997), no. 2, 209–270.

[HMS74] D. Husemoller, J. C. Moore, and J. Stasheff, *Differential homological algebra and homogeneous spaces*, J. Pure Appl. Algebra **5** (1974), 113–185.

[Hop87] M. J. Hopkins, *Global methods in homotopy theory*, Homotopy theory (Durham, 1985) (J. D. S. Jones and E. Rees, eds.), Cambridge Univ. Press, Cambridge-New York, 1987, LMS Lecture Note Series 117, pp. 73–96.

[Hov99] M. Hovey, *Model categories*, American Mathematical Society, Providence, RI, 1999.

[HPa] M. Hovey and J. H. Palmieri, *Galois theory of thick subcategories in modular representation theory*, preprint.

[HPb] M. Hovey and J. H. Palmieri, *Stably thick subcategories of modules over Hopf algebras*, preprint.

[HP93] M. J. Hopkins and J. H. Palmieri, *A nilpotence theorem for modules over the mod 2 Steenrod algebra*, Topology **32** (1993), no. 4, 751–756.

[HPS97] M. Hovey, J. H. Palmieri, and N. P. Strickland, *Axiomatic stable homotopy theory*, Mem. Amer. Math. Soc. **128** (1997), no. 610, x+114.

[HPS99] M. J. Hopkins, J. H. Palmieri, and J. H. Smith, *Vanishing lines in generalized Adams spectral sequences are generic*, Geom. Topol. **3** (1999), 155–165.

[HS98] M. J. Hopkins and J. H. Smith, *Nilpotence and stable homotopy theory II*, Ann. of Math. (2) **148** (1998), no. 1, 1–49.

[JK81] G. James and A. Kerber, *The representation theory of the symmetric group*, Addison-Wesley Publishing Co., Reading, Mass., 1981, with a foreword by P. M. Cohn, with an introduction by Gilbert de B. Robinson.

[Lin77a] W. H. Lin, *Cohomology of sub-Hopf algebras of the Steenrod algebra*, J. Pure Appl. Algebra **10** (1977), 101–114.

[Lin77b] W. H. Lin, *Cohomology of sub-Hopf algebras of the Steenrod algebra II*, J. Pure Appl. Algebra **11** (1977), 105–110.

[LS69] R. G. Larson and M. E. Sweedler, *An associative orthogonal bilinear form for Hopf algebras*, Amer. J. Math. **91** (1969), 75–94.

[Mah70] M. E. Mahowald, *The order of the image of the J-homomorphism*, Bull. Amer. Math. Soc. **76** (1970), 1310–1313.

[Mar] H. R. Margolis, *Unpublished notes*.

[Mar83] H. R. Margolis, *Spectra and the Steenrod algebra*, North-Holland Publishing Co., Amsterdam-New York, 1983, Modules over the Steenrod algebra and the stable homotopy category.

[May69] J. P. May, *Matric Massey products*, J. Algebra **12** (1969), 533–568.

[May70] J. P. May, *A general algebraic approach to Steenrod operations*, The Steenrod Algebra and its Applications (Proc. Conf. to Celebrate N. E. Steenrod's Sixtieth Birthday, Battelle Memorial Inst., Columbus, Ohio, 1970) (F. P. Peterson, ed.), Springer, Berlin, 1970, Lecture Notes in Mathematics, Vol. 168, pp. 153–231.

[Mil58] J. W. Milnor, *The Steenrod algebra and its dual*, Ann. of Math. (2) **67** (1958), 150–171.

[Mil78] H. R. Miller, *A localization theorem in homological algebra*, Math. Proc. Cambridge Philos. Soc. **84** (1978), no. 1, 73–84.

[Mil81] H. R. Miller, *On relations between Adams spectral sequences, with an application to the stable homotopy of a Moore space*, J. Pure Appl. Algebra **20** (1981), no. 3, 287–312.

[Mil92] H. R. Miller, *Finite localizations*, Bol. Soc. Mat. Mexicana (2) **37** (1992), no. 1-2, 383–389, Papers in honor of José Adem (Spanish).

[Mit85] S. Mitchell, *Finite complexes with $A(n)$-free cohomology*, Topology **24** (1985), no. 2, 227–246.

[Mit86] S. Mitchell, *On the Steinberg module, representations of the symmetric groups, and the Steenrod algebra*, J. Pure Appl. Algebra **39** (1986), no. 3, 275–281.

[MM65] J. W. Milnor and J. C. Moore, *On the structure of Hopf algebras*, Ann. of Math. (2) **81** (1965), 211–264.

[MP72] J. C. Moore and F. P. Peterson, *Modules over the Steenrod algebra*, Topology **11** (1972), 387–395.

[MPT71] H. R. Margolis, S. B. Priddy, and M. C. Tangora, *Another systematic phenomenon in the cohomology of the Steenrod algebra*, Topology **10** (1971), 43–46.

[MR] M. E. Mahowald and C. Rezk, *Brown-Comenetz duality and the Adams spectral sequence*, preprint.

[MS87] M. E. Mahowald and P. L. Shick, *Periodic phenomena in the classical Adams spectral sequence*, Trans. Amer. Math. Soc. **300** (1987), no. 1, 191–206.

[MS95] M. E. Mahowald and H. Sadofsky, v_n-*telescopes and the Adams spectral sequence*, Duke Math. J. **78** (1995), no. 1, 101–129.

[MT68] M. E. Mahowald and M. C. Tangora, *An infinite subalgebra of* $\operatorname{Ext}_A(Z_2, Z_2)$, Trans. Amer. Math. Soc. **132** (1968), 263–274.

[MW81] H. R. Miller and C. Wilkerson, *Vanishing lines for modules over the Steenrod algebra*, J. Pure Appl. Algebra **22** (1981), no. 3, 293–307.

[Nak75] O. Nakamura, *Some differentials in the mod 3 Adams spectral sequence*, Bull. Sci. Engrg. Div. Univ. Ryukyus Math. Natur. Sci. (1975), no. 19, 1–25.

[Nis73] G. Nishida, *The nilpotency of elements of the stable homotopy groups of spheres*, J. Math. Soc. Japan **25** (1973), 707–732.

[NP98] D. K. Nakano and J. H. Palmieri, *Support varieties for the Steenrod algebra*, Math. Z. **227** (1998), no. 4, 663–684.

[Ohk89] T. Ohkawa, *The injective hull of homotopy types with respect to generalized homology functors*, Hiroshima Math. J. **19** (1989), no. 3, 631–639.

[Pal91] J. H. Palmieri, *A chromatic spectral sequence to study Ext over the Steenrod algebra*, Ph.D. thesis, Mass. Inst. of Tech., 1991.

[Pal92] J. H. Palmieri, *Self-maps of modules over the Steenrod algebra*, J. Pure Appl. Algebra **79** (1992), no. 3, 281–291.

[Pal94] J. H. Palmieri, *The chromatic filtration and the Steenrod algebra*, Topology and representation theory (Evanston, IL, 1992) (E. Friedlander and M. E. Mahowald, eds.), Contemp. Math., vol. 158, Amer. Math. Soc., Providence, RI, 1994, pp. 187–201.

[Pal96a] J. H. Palmieri, *Nilpotence for modules over the mod 2 Steenrod algebra I*, Duke Math. J. **82** (1996), 195–208.

[Pal96b] J. H. Palmieri, *Nilpotence for modules over the mod 2 Steenrod algebra II*, Duke Math. J. **82** (1996), 209–226.

[Pal97] J. H. Palmieri, *A note on the cohomology of finite dimensional cocommutative Hopf algebras*, J. Algebra **188** (1997), no. 1, 203–215.

[Pal99] J. H. Palmieri, *Quillen stratification for the Steenrod algebra*, Ann. of Math. (2) **149** (1999), 421–449.

[PS94] J. H. Palmieri and H. Sadofsky, *Self maps of spectra, a theorem of J. Smith, and Margolis' killing construction*, Math. Z. **215** (1994), no. 3, 477–490.

[Qui67] D. G. Quillen, *Homotopical algebra*, Lecture Notes in Mathematics, vol. 43, Springer-Verlag, Berlin, 1967.

[Qui71] D. G. Quillen, *The spectrum of an equivariant cohomology ring I, II*, Ann. of Math. (2) **94** (1971), 549–572, 573–602.

[Rad77] D. Radford, *Pointed Hopf algebras are free over Hopf subalgebras*, J. Algebra **45** (1977), no. 2, 266–273.

[Rav84] D. C. Ravenel, *Localization with respect to certain periodic homology theories*, Amer. J. Math. **106** (1984), no. 2, 351–414.

[Rav86] D. C. Ravenel, *Complex cobordism and stable homotopy groups of spheres*, Academic Press, Inc., Orlando, Fla., 1986.

[Rav92] D. C. Ravenel, *Nilpotence and periodicity in stable homotopy theory*, Annals of Mathematics Studies, vol. 128, Princeton University Press, Princeton, NJ, 1992, Appendix C by Jeff Smith.

[Rot77] M. Roth, Ph.D. thesis, Johns Hopkins University, 1977.

[Saw82] J. Sawka, *Odd primary Steenrod operations in first quadrant spectral sequences*, Trans. Amer. Math. Soc. **273** (1982), no. 2, 737–752.

[Shi88] P. L. Shick, *Odd primary periodic phenomena in the classical Adams spectral sequence*, Trans. Amer. Math. Soc. **309** (1988), no. 1, 77–86.

[Sin73] W. M. Singer, *Steenrod squares in spectral sequences I, II*, Trans. Amer. Math. Soc. **175** (1973), 327–336, 337–353.

[Sin99] W. M. Singer, *Steenrod squares in spectral sequences: The cohomology of Hopf algebra extensions and classifying spaces*, preprint, 1999.

[Smi] J. H. Smith, *Finite complexes with vanishing lines of small slopes*, preprint.

[Swe69] M. E. Sweedler, *Hopf algebras*, W. A. Benjamin, 1969.

[Wei94] C. A. Weibel, *An introduction to homological algebra*, Cambridge studies in advanced mathematics, vol. 38, Cambridge University Press, 1994.

[Wil81] C. Wilkerson, *The cohomology algebras of finite dimensional Hopf algebras*, Trans. Amer. Math. Soc. **264** (1981), no. 1, 137–150.

[Zac67] A. Zachariou, *A subalgebra of* $\operatorname{Ext}_A^{**}(Z_2, Z_2)$, Bull. Amer. Math. Soc. **73** (1967), 647–648.

Index

$\frac{1}{2}A(n)$, 85–99

A, 36
act nilpotently, *see* nilpotent action
Adams spectral sequence, *see* spectral sequence, Adams
Adams tower, 25
admissible, 147–150
algebra, 2
$A(n)$, 38, 50, 54, 58, 71, 74, 85–99
    A-comodule structure of, 85–99
a_n, 80
$A(n)_k$, 91–95
antipode, 2, 3
augmentation, 2

bicommutative, 3
bounded below, 1
Bousfield class, 29, 61, 120, 139
Bousfield equivalence, 29, 120, 139
BP, 108
$\beta\widetilde{\mathscr{P}}^0$, 21, 123
$\beta\widetilde{\mathscr{P}}^n$, 155
Brown category, 15
Brown-Comenetz duality, *see* dual, Brown-Comenetz
b_{ts}, 38, 55, 57, 58, 92, 124, 157
    nilpotence of, 123, 157–163

category
    derived, 15
    model, 15, 153–154
        cofibrantly generated, 154
    stable homotopy, 14
        connective, 22
cellular tower, 22
    strong, 22
change-of-rings isomorphism, 18
Chouinard's theorem, 31, 124
chromatic convergence, 75–76
chromatic tower, 75
CL, 24, 49, 58, 68
closed model category, *see* category, model
coaction, 4, 20, 109, 121, 135, 137
    diagonal, 4
coalgebra, 2
    quotient, 16, 18, 19, 59

coaugmentation, 2
cobar complex, 8, *see* complex, cobar
cocommutative, 3
cofiber sequence, 14, 16, 21, 30
cofibration, 154, *see also* cofiber sequence
cofree, 6, 7, 72, 87
cohomology functor, *see* functor, cohomology
coideal
    coaugmentation, 20
colimit
    homotopy, 15
commutative, 3
comodule, 4
    simple, 14, 22, 31, 153
comodule-like, *see* CL
complete, 26, 80
complex
    cobar, 8
comultiplication, 2
conjugation, 2, 3, 135–137, 140
connective, 23, 25
    weakly, 23
conormal, *see* Hopf algebra, quotient, conormal
coproduct, 15
cotensor product, 16
counit, 2

D, 108, 112
degree, 1
derived category, *see* category, derived
detect nilpotence, 110–112, 139
diagonal, 2
$D(n)$, 112, 132
D_r, 119
$D_{r,q}$, 119
dual
    Brown-Comenetz, 30, 125, 139
    graded, 3, 5
    module, 6, 140
    Spanier-Whitehead, 48, 60, 129
DX, *see* dual, Spanier-Whitehead
$D[x]$, 9
    classification of comodules over, 10
    Ext over, 11
$D[\xi_t^{p^s}]$, 37, 55

169

E-complete, *see* complete
elementary, *see* Hopf algebra, elementary
e_n, 97–99
$E[\tau_n]$, 37, 55
$E[x]$, 9
    classification of comodules over, 10
    Ext over, 11
Ext, 8, 15
Ext_A, 49
extension, *see* Hopf algebra extension; group extension

F-isomorphism, 109, 110, 122, 128, 134, 135, 143
    uniform, 109
fibration, 154
field spectrum, *see* spectrum, field
finite spectrum, *see* spectrum, finite
finite type, 1, 27, 30, 140
finitely generated
    essentially, 144–146
flat, 25, 26, 80, 101
$F(u_{d_1}^{j_1}, \ldots, u_{d_m}^{j_m})$, 60–63, 66–71, 91
$F(U_n)$, *see* $F(u_{d_1}^{j_1}, \ldots, u_{d_m}^{j_m})$
functor
    cohomology, 16
    exact, 72
    homology, 16
    pth power, 94, 149

$\Gamma \square_B k$, 20
$\Gamma // C$, 20
generated, 14
generic, 24, 26, 114, 128, 129
grading, 1
group, 3
group algebra, 3, 31
group extension, 31
grouplike, 4

H, 16–22, 50
    Tate version of, 71–75
$\widehat{H}$, *see* H, Tate version of
HB-complete, *see* complete
HD, 108, 110, 112, 122, 128, 135, 146, 151
Hom, 72
homology functor, *see* functor, homology
homotopy group, 16
hook length, 97
Hopf algebra, 2, 153
    connected, 3, 23
    elementary, 39, 41, 53, 57
    quasi-elementary, 41, 108, 109, 111, 121, 139
    quotient, 16, 19, 36, 51, 108, 139, 140
        conormal, 20, 27, 36, 109, 112, 121, 139
        maximal elementary, 40
        maximal quasi-elementary, 119, 134, 137
    sub-
        normal, 20
Hopf algebra extension, 20, 30, 37, 51, 52, 55, 56, 112, 139, 140
horizontal vanishing line, *see* vanishing line, horizontal
h_{ts}, 38, 51, 57, 58, 92, 115, 135, 137, 156, 157
    nilpotence of, 51, 116, 120, 123, 156
Hurewicz map, 23, 58, 109, 112
Hurewicz theorem, 23
hypercohomology, 16, 29

ideal, 127, 131–132
    augmentation, 20
    invariant, 127, 131, 144, 146, 151
idempotent, 96–99
induction, 17, 71
injective, 6–8, 17–20, 44, 59, 72
intercept, 50
invariants, 108, 134
isomorphism
    change-of-rings, 18
IX, *see* dual, Brown-Comenetz
$I(X)$, *see* ideal

kG, 31
$K(n)$, *see* Morava K-theories
Künneth isomorphism, 8, 20

$L_{\mathscr{C}}^f$, *see* localization, finite
L_n^f, *see* localization, finite
Lie algebra, 3
    restricted, 4
limit
    homotopy, 15
L_n^f, 73
$L_n^f A(m)$, 74
$\text{loc}(A)$, 70, 142
$\text{loc}(\Gamma)$, 24
$\text{loc}(Y)$, 24
localization
    Bousfield, 66, 71
    finite, 65, 66, 72, 75
    smashing, 66, 71
localizing subcategory, *see* subcategory, localizing
locally finite, 6

Margolis' killing construction, 65, 68
Massey product, 21, 156, 157, 159–163
$M\langle n+1, \infty \rangle$, 70
M_n^f, 75
module, 4
monogenic, 14, 22, 29
Moore spectrum, 104–106
Morava K-theories, 57, 128
morphism, 14, 72
    grading, 14

stable, 72
multiplication, 2

nilpotence theorem, 110, 111
nilpotent action, 51, 56
Nishida's theorem, 110
normal, *see* Hopf algebra, sub-, normal

$\widetilde{\mathscr{P}}$, 155
P-algebra, 38
periodicity theorem, 128, 129
Φ, 94, 149
π_{**}, 16, 49
PM, 108
$P(n)$, 86–88, 90–99
$P(n)_k$, 91–95
Postnikov tower, 23
primitive, 4, 8, 21, 59, 108, 123, 154, 156–158
profile function, 37
projective, 6
P_t^s, 42, 53, 57, 146
p_t^s, 42, 148
P_t^s-homology, 42, 43, 128
  module, 42, 45
P_t^s-homology spectrum, 42, 45, 147

$\mathscr{Q}$, 109, 121, 134
$\overline{\mathscr{Q}}$, 121
Q_n, 42, 43, 57, 80–85, 147
q_n, 43, 80–81, 88–90
q_n-complete, *see* complete
Q_n-homology, 43
Q_n-homology spectrum, 43
$(Q_n)_{**}(Q_n)$, 88, 101–102
$(q_n)_{**}(q_n)$, 99
quasi-elementary, *see* Hopf algebra, quasi-elementary
Quillen stratification, 109, 124

rank variety, *see* variety, rank
relative vanishing line, *see* vanishing line, relative
resolution
  injective, 8, 10, 153
restriction, 4, 17, 71
restriction map, 19, 58, 112
ring spectrum, *see* spectrum, ring

self-map, 57, 110, 111
  central, 59, 128
Shapiro's lemma, 18
shearing isomorphism, 18
Σ_n, *see* symmetric group
$\Sigma^{i,j}$, *see* suspension
slope, 46, 49, 57, 60, 66, 73, 75, 147–150
slope support, 147–150
Slopes, 147–150
Slopes', 147–150

small, 24
smash product, 15
Spanier-Whitehead dual, *see* dual, Spanier-Whitehead
spectral sequence
  Adams, 25–29, 80–85, 114
    construction, 25
    convergence, 26
  Atiyah-Hirzebruch, 57, 144
  change-of-rings, *see* spectral sequence, extension
  extension, 28, 112, 113
  Lyndon-Hochschild-Serre, 28
spectrum, 14
  field, 19, 45, 80, 81, 128
  finite, 24, 63, 65, 67, 110, 111, 128, 142, 143, 146–148
  ring, 16, 19, 25, 45, 110, 111, 128, 129
sphere, 15
Sq^n, 137
$\widetilde{\text{Sq}}^n$, 155, 156
stable homotopy category, *see* category, stable homotopy
Stable(A), 36
Stable(Γ), 14, 153
StComod(B), 72
Steenrod algebra, 4, 15, 36
Steenrod operation, 21, 29, 155–156
subcategory
  localizing, 24, 49, 66
  thick, 24, 26, 61, 65, 72, 90–95, 116, 129, 143–146
suspension, 1, 14, 72
  internal, 1, 10
symmetric group, 96–99

tame, 6
telescope, 30, 63, 66
telescope conjecture, 71
tensor product, 4
thick subcategory, *see* subcategory, thick
thick(Y), 24
τ_n, 36
trivial comodule, 22
twist map, 2
type n, 57, 63, 67

u_n, 57, 59, 149
u_n-map, 57, 60–62, 66
unit, 2
unit map, 16
universal enveloping algebra, 3

$v_0^{-1}\pi_{**}$, 101
vanishing line, 49–60, 70
  horizontal, 49, 53, 57
  relative, 58
vanishing plane, 26–27, 83, 88–90, 114
variety

rank, 146–147
$V(n)$, 86–88
v_n, 38, 55, 57, 59, 80–85, 92, 157
v_n-map, 57, 81, 128

weak equivalence, 154
wedge, 15
W_j, 114, 115

$\langle X \rangle$, *see* Bousfield class
$[x]$, 21, 156
$|x|$, *see* degree
X-acyclic, 29
ξ_n, 36

$[y]$, *see* $[x]$
y-element, *see* y-map
y-map, 128–131, 146
    weak, 129
$Y(n)$, 102–104
y_n, 46, 59, 73
Young diagram, 97–99
Young tableau, 97–99

Z_n, 96–99
$Z(n)$, 46, 49, 57, 60–62, 66, 68, 75
$z(n)$, 46, 57, 59
ζ_n, 135

Editorial Information

To be published in the *Memoirs*, a paper must be correct, new, nontrivial, and significant. Further, it must be well written and of interest to a substantial number of mathematicians. Piecemeal results, such as an inconclusive step toward an unproved major theorem or a minor variation on a known result, are in general not acceptable for publication. Papers appearing in *Memoirs* are generally longer than those appearing in *Transactions*, which shares the same editorial committee.

As of January 31, 2001, the backlog for this journal was approximately 7 volumes. This estimate is the result of dividing the number of manuscripts for this journal in the Providence office that have not yet gone to the printer on the above date by the average number of monographs per volume over the previous twelve months, reduced by the number of volumes published in four months (the time necessary for preparing a volume for the printer). (There are 6 volumes per year, each containing at least 4 numbers.)

A Consent to Publish and Copyright Agreement is required before a paper will be published in the *Memoirs*. After a paper is accepted for publication, the Providence office will send a Consent to Publish and Copyright Agreement to all authors of the paper. By submitting a paper to the *Memoirs*, authors certify that the results have not been submitted to nor are they under consideration for publication by another journal, conference proceedings, or similar publication.

Information for Authors

Memoirs are printed from camera copy fully prepared by the author. This means that the finished book will look exactly like the copy submitted.

The paper must contain a *descriptive title* and an *abstract* that summarizes the article in language suitable for workers in the general field (algebra, analysis, etc.). The *descriptive title* should be short, but informative; useless or vague phrases such as "some remarks about" or "concerning" should be avoided. The *abstract* should be at least one complete sentence, and at most 300 words. Included with the footnotes to the paper should be the 2000 *Mathematics Subject Classification* representing the primary and secondary subjects of the article. The classifications are accessible from www.ams.org/msc/. The list of classifications is also available in print starting with the 1999 annual index of *Mathematical Reviews*. The Mathematics Subject Classification footnote may be followed by a list of *key words and phrases* describing the subject matter of the article and taken from it. Journal abbreviations used in bibliographies are listed in the latest *Mathematical Reviews* annual index. The series abbreviations are also accessible from www.ams.org/publications/. To help in preparing and verifying references, the AMS offers MR Lookup, a Reference Tool for Linking, at www.ams.org/mrlookup/. When the manuscript is submitted, authors should supply the editor with electronic addresses if available. These will be printed after the postal address at the end of the article.

Electronically prepared manuscripts. The AMS encourages electronically prepared manuscripts, with a strong preference for $\mathcal{AMS}$-LaTeX. To this end, the Society has prepared $\mathcal{AMS}$-LaTeX author packages for each AMS publication. Author packages include instructions for preparing electronic manuscripts, the *AMS Author Handbook*, samples, and a style file that generates the particular design specifications of that publication series. Though $\mathcal{AMS}$-LaTeX is the highly preferred format of TeX, author packages are also available in $\mathcal{AMS}$-TeX.

Authors may retrieve an author package from e-MATH starting from `www.ams.org/tex/` or via FTP to `ftp.ams.org` (login as `anonymous`, enter username as password, and type `cd pub/author-info`). The *AMS Author Handbook* and the *Instruction Manual* are available in PDF format following the author packages link from `www.ams.org/tex/`. The author package can be obtained free of charge by sending email to `pub@ams.org` (Internet) or from the Publication Division, American Mathematical Society, P.O. Box 6248, Providence, RI 02940-6248. When requesting an author package, please specify $\mathcal{AMS}$-LaTeX or $\mathcal{AMS}$-TeX, Macintosh or IBM (3.5) format, and the publication in which your paper will appear. Please be sure to include your complete mailing address.

Sending electronic files. After acceptance, the source file(s) should be sent to the Providence office (this includes any TeX source file, any graphics files, and the DVI or PostScript file).

Before sending the source file, be sure you have proofread your paper carefully. The files you send must be the EXACT files used to generate the proof copy that was accepted for publication. For all publications, authors are required to send a printed copy of their paper, which exactly matches the copy approved for publication, along with any graphics that will appear in the paper.

TeX files may be submitted by email, FTP, or on diskette. The DVI file(s) and PostScript files should be submitted only by FTP or on diskette unless they are encoded properly to submit through email. (DVI files are binary and PostScript files tend to be very large.)

Electronically prepared manuscripts can be sent via email to `pub-submit@ams.org` (Internet). The subject line of the message should include the publication code to identify it as a Memoir. TeX source files, DVI files, and PostScript files can be transferred over the Internet by FTP to the Internet node `e-math.ams.org` (130.44.1.100).

Electronic graphics. Comprehensive instructions on preparing graphics are available at `www.ams.org/jourhtml/graphics.html`. A few of the major requirements are given here.

Submit files for graphics as EPS (Encapsulated PostScript) files. This includes graphics originated via a graphics application as well as scanned photographs or other computer-generated images. If this is not possible, TIFF files are acceptable as long as they can be opened in Adobe Photoshop or Illustrator. No matter what method was used to produce the graphic, it is necessary to provide a paper copy to the AMS.

Authors using graphics packages for the creation of electronic art should also avoid the use of any lines thinner than 0.5 points in width. Many graphics packages allow the user to specify a "hairline" for a very thin line. Hairlines often look acceptable when proofed on a typical laser printer. However, when produced on a high-resolution laser imagesetter, hairlines become nearly invisible and will be lost entirely in the final printing process.

Screens should be set to values between 15% and 85%. Screens which fall outside of this range are too light or too dark to print correctly. Variations of screens within a graphic should be no less than 10%.

Inquiries. Any inquiries concerning a paper that has been accepted for publication should be sent directly to the Electronic Prepress Department, American Mathematical Society, P. O. Box 6248, Providence, RI 02940-6248.

Editors

This journal is designed particularly for long research papers, normally at least 80 pages in length, and groups of cognate papers in pure and applied mathematics. Papers intended for publication in the *Memoirs* should be addressed to one of the following editors. In principle the Memoirs welcomes electronic submissions, and some of the editors, those whose names appear below with an asterisk (*), have indicated that they prefer them. However, editors reserve the right to request hard copies after papers have been submitted electronically. Authors are advised to make preliminary email inquiries to editors about whether they are likely to be able to handle submissions in a particular electronic form.

Algebra to CHARLES CURTIS, Department of Mathematics, University of Oregon, Eugene, OR 97403-1222 email: `cwc@darkwing.uoregon.edu`

Algebraic geometry and commutative algebra to LAWRENCE EIN, Department of Mathematics, University of Illinois, 851 S. Morgan (M/C 249), Chicago, IL 60607-7045; email: `ein@uic.edu`

Algebraic topology and cohomology of groups to STEWART PRIDDY, Department of Mathematics, Northwestern University, 2033 Sheridan Road, Evanston, IL 60208-2730; email: `priddy@math.nwu.edu`

Combinatorics and Lie theory to SERGEY FOMIN, Department of Mathematics, University of Michigan, Ann Arbor, Michigan 48109-1109; email: `fomin@math.lsa.umich.edu`

Complex analysis and complex geometry to DUONG H. PHONG, Department of Mathematics, Columbia University, 2990 Broadway, New York, NY 10027-0029; email: `dp@math.columbia.edu`

*__Differential geometry and global analysis__ to LISA C. JEFFREY, Department of Mathematics, University of Toronto, 100 St. George St., Toronto, ON Canada M5S 3G3; email: `jeffrey@math.toronto.edu`

*__Dynamical systems and ergodic theory__ to ROBERT F. WILLIAMS, Department of Mathematics, University of Texas, Austin, Texas 78712-1082; email: `bob@math.utexas.edu`

Geometric topology, knot theory and hyperbolic geometry to ABIGAIL A. THOMPSON, Department of Mathematics, University of California, Davis, Davis, CA 95616-5224; email: `thompson@math.ucdavis.edu`

Harmonic analysis, representation theory, and Lie theory to ROBERT J. STANTON, Department of Mathematics, The Ohio State University, 231 West 18th Avenue, Columbus, OH 43210-1174; email: `stanton@math.ohio-state.edu`

*__Logic__ to THEODORE SLAMAN, Department of Mathematics, University of California, Berkeley, CA 94720-3840; email: `slaman@math.berkeley.edu`

Number theory to MICHAEL J. LARSEN, Department of Mathematics, Indiana University, Bloomington, IN 47405; email: `larsen@math.indiana.edu`

Operator algebras and functional analysis to BRUCE E. BLACKADAR, Department of Mathematics, University of Nevada, Reno, NV 89557; email: `bruceb@math.unr.edu`

*__Ordinary differential equations, partial differential equations, and applied mathematics__ to PETER W. BATES, Department of Mathematics, Brigham Young University, 292 TMCB, Provo, UT 84602-1001; email: `peter@math.byu.edu`

*__Partial differential equations and applied mathematics__ to BARBARA LEE KEYFITZ, Department of Mathematics, University of Houston, 4800 Calhoun Road, Houston, TX 77204-3476; email: `keyfitz@uh.edu`

*__Probability and statistics__ to KRZYSZTOF BURDZY, Department of Mathematics, University of Washington, Box 354350, Seattle, Washington 98195-4350; email: `burdzy@math.washington.edu`

*__Real and harmonic analysis and geometric partial differential equations__ to WILLIAM BECKNER, Department of Mathematics, University of Texas, Austin, TX 78712-1082; email: `beckner@math.utexas.edu`

All other communications to the editors should be addressed to the Managing Editor, WILLIAM BECKNER, Department of Mathematics, University of Texas, Austin, TX 78712-1082; email: `beckner@math.utexas.edu`.

Selected Titles in This Series

(Continued from the front of this publication)

686 **L. Gaunce Lewis, Jr.,** Splitting theorems for certain equivariant spectra, 2000

685 **Jean-Luc Joly, Guy Metivier, and Jeffrey Rauch,** Caustics for dissipative semilinear oscillations, 2000

684 **Harvey I. Blau, Bangteng Xu, Z. Arad, E. Fisman, V. Miloslavsky, and M. Muzychuk,** Homogeneous integral table algebras of degree three: A trilogy, 2000

683 **Serge Bouc,** Non-additive exact functors and tensor induction for Mackey functors, 2000

682 **Martin Majewski,** ational homotopical models and uniqueness, 2000

681 **David P. Blecher, Paul S. Muhly, and Vern I. Paulsen,** Categories of operator modules (Morita equivalence and projective modules, 2000

680 **Joachim Zacharias,** Continuous tensor products and Arveson's spectral C^*-algebras, 2000

679 **Y. A. Abramovich and A. K. Kitover,** Inverses of disjointness preserving operators, 2000

678 **Wilhelm Stannat,** The theory of generalized Dirichlet forms and its applications in analysis and stochastics, 1999

677 **Volodymyr V. Lyubashenko,** Squared Hopf algebras, 1999

676 **S. Strelitz,** Asymptotics for solutions of linear differential equations having turning points with applications, 1999

675 **Michael B. Marcus and Jay Rosen,** Renormalized self-intersection local times and Wick power chaos processes, 1999

674 **R. Lawther and D. M. Testerman,** A_1 subgroups of exceptional algebraic groups, 1999

673 **John Lott,** Diffeomorphisms and noncommutative analytic torsion, 1999

672 **Yael Karshon,** Periodic Hamiltonian flows on four dimensional manifolds, 1999

671 **Andrzej Rosłanowski and Saharon Shelah,** Norms on possibilities I: Forcing with trees and creatures, 1999

670 **Steve Jackson,** A computation of δ_5^1, 1999

669 **Seán Keel and James McKernan,** Rational curves on quasi-projective surfaces, 1999

668 **E. N. Dancer and P. Poláčik,** Realization of vector fields and dynamics of spatially homogeneous parabolic equations, 1999

667 **Ethan Akin,** Simplicial dynamical systems, 1999

666 **Mark Hovey and Neil P. Strickland,** Morava K-theories and localisation, 1999

665 **George Lawrence Ashline,** The defect relation of meromorphic maps on parabolic manifolds, 1999

664 **Xia Chen,** Limit theorems for functionals of ergodic Markov chains with general state space, 1999

663 **Ola Bratteli and Palle E. T. Jorgensen,** Iterated function systems and permutation representation of the Cuntz algebra, 1999

662 **B. H. Bowditch,** Treelike structures arising from continua and convergence groups, 1999

661 **J. P. C. Greenlees,** Rational S^1-equivariant stable homotopy theory, 1999

660 **Dale E. Alspach,** Tensor products and independent sums of $\mathcal{L}_p$-spaces, $1 < p < \infty$, 1999

659 **R. D. Nussbaum and S. M. Verduyn Lunel,** Generalizations of the Perron-Frobenius theorem for nonlinear maps, 1999

658 **Hasna Riahi,** Study of the critical points at infinity arising from the failure of the Palais-Smale condition for n-body type problems, 1999

657 **Richard F. Bass and Krzysztof Burdzy,** Cutting Brownian paths, 1999

For a complete list of titles in this series, visit the
AMS Bookstore at **www.ams.org/bookstore/**.